LIQUID PHASE SINTERING

LIQUID PHASE SINTERING

RANDALL M. GERMAN
Rensselaer Polytechnic Institute
Troy, New York

Springer Science+Business Media, LLC

Library of Congress Cataloging in Publication Data

German, R. M.
Liquid phase sintering.

Includes bibliographies and index.
1. Sintering. 2. Phase rule and equilibrium. I. Title.
TN695.G469 1985 671.3′73 85-25786

ISBN 978-1-4899-3601-1 ISBN 978-1-4899-3599-1 (eBook)
DOI 10.1007/978-1-4899-3599-1

This limited facsimile edition has been issued for the purpose of keeping this title available to the scientific community.

Originally published by Plenum Press New York in 1985.
Softcover reprint of the hardcover 1st edition 1985

Dedicated to

PROFESSOR FRITZ V. LENEL

a pioneer in the field of liquid phase sintering

FOREWORD

In the past few years there has been rapid growth in the activities involving particulate materials because of recognized advantages in manufacturing. This growth is attributed to several factors; i) an increased concern over energy utilization, ii) a desire to better control microstructure in engineering materials, iii) the need for improved material economy, iv) societal and economic pressures for higher productivity and quality, v) requirements for unique property combinations for high performance applications, and vi) a desire for net shape forming. Accordingly, liquid phase sintering has received increased attention as part of the growth in particulate materials processing. As a consequence, the commercial applications for liquid phase sintering are expanding rapidly. This active and expanding interest is not well served by available texts. For this reason I felt it was appropriate to write this book on liquid phase sintering.

The technology of liquid phase sintering is quite old and has been in use in the ceramics industry for many centuries. However, the general perception among materials and manufacturing engineers is that liquid phase sintering is still a novel technique. I believe the diverse technological applications outlined in this book will dispell such impressions. Liquid phase sintering has great value in fabricating several unique materials to near net shapes and will continue to expand in applications as the fundamental attributes are better appreciated.

I am personally involved with several uses for liquid phase sintering. My initial exposure was with refractory metals and alloys like the heavy alloys based on high tungsten contents. My subsequent studies have included ferrous structural alloys, aluminum alloys, cemented carbides, molybdenum, metal-ceramic composites, alumina, and several high performance ceramics processed by liquid phase sintering. Obviously, my early emphasis on the metallic examples of liquid phase sintering are carried into this book; however, I have made an effort to include several examples of nonmetallic materials.

In writing the book, my goal was to organize the current knowledge on liquid phase sintering into a general, material independent treatment. The primary emphasis is on the fundamentals and the universal characteristics of liquid phase sintering. Through an understanding of the fundamentals, the technology for manufacturing by liquid phase sintering becomes evident. The book is written for the technical audience interested in liquid phase sintering, including manufacturing and materials engineers, graduate students, research scientists, development engineers, and university professors. It is organized into four major divisions. Chapters 1, 2, and 3 give an introduction to the basic phenomena underlying liquid phase sintering. Chapters 4, 5, and 6 treat the classic process of persistent liquid phase sintering. Next, Chapter 7 introduces the special techniques for consolidation of powders with liquid phases. Finally, Chapters 8, 9, and 10 contain an overview of the technological factors, including the processing variables, properties, and some applications. Chapter 10, the discussion of applications, proved to be more difficult to write than anticipated. Certain applications are covered heavily in the literature, while there is negligible documentation on other important applications. Hence, only a survey of the applications is provided.

It is evident that several of the topics could have been treated in greater detail. However, the concept for this book was to organize the information without delving into the details appropriate to review journals. I hope the reader appreciates the need for brevity in light of the broad spectrum of topics appropriate to the subject.

My efforts were aided by several people. The comments of both Barry Rabin and Sandra Hillman were of great value in revising the early drafts. In addition, Joseph Strauss, Animesh Bose, and Donald Polensky were helpful in making comments on a preliminary draft of the book. I especially wish to thank Professor Fritz Lenel for his extensive help and several comments on the manuscript; it is to him that I dedicate this book.

Randall M. German   Troy, New York   August 1985

CONTENTS

CHAPTER ONE

Introduction to Liquid Phase Sintering

A. Initial Definitions

Packed powders will bond together when heated to temperatures in excess of approximately half of the absolute melting temperature. This phenomenon is termed sintering. A common characteristic of all forms of sintering is a reduction in surface area with concomitant compact strengthening. This occurs through the formation of interparticle bonds brought about by atomic motion at the sintering temperature. During liquid phase sintering a liquid phase coexists with a particulate solid at the sintering temperature. The liquid phase usually enhances the rate of interparticle bonding during sintering. Accompanying interparticle bonding are significant changes in the pore structure and compact properties including strength, ductility, conductivity, magnetic permeability, and corrosion resistance.

Solid state sintering of single component materials is the best understood form of sintering. Even for this case there is a complexity of steps through which the powder progresses as it is heated. Computer routines have been developed to model the complex events, based on basic material constants like diffusivity. For simple systems involving ideal conditions, these calculations have verified our basic understanding of sintering. However, even in solid state sintering various complications exist which degrade the ability to predict the progression and properties of sintered compacts. One example of such a factor is the particle size distribution. Most theories assume a monosized spherical powder, while most practical sintering involves particles in a range of sizes, and often far from spherical in shape.

In a general categorization of sintering techniques, pressure is the first consideration. Most sintering is performed without an external pressure (pressureless sintering). For many high-performance applications, high densities are attained using external pressure sources. Such techniques as hot pressing, hot isostatic pressing, hot forging, and hot extrusion use a combination of temperature, stress, and strain rate to densify powder compacts. For the most part, liquid phase sintering exhibits sufficient internal force through liquid capillary action on the particulate solid that external forces are not required. The magnitude of the capillary force is equivalent to very large external pressures.

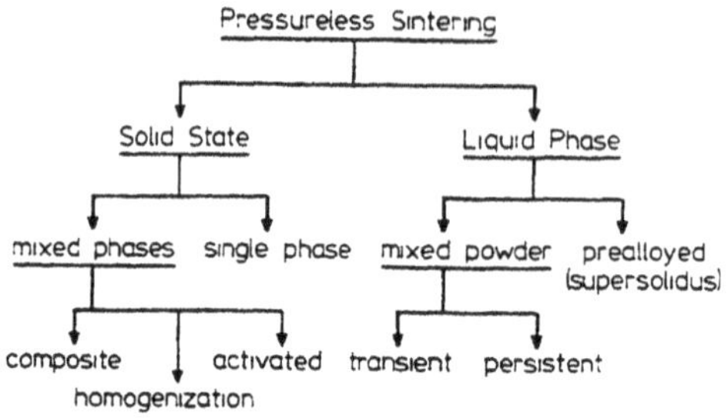

Figure 1.1 The subdivisions of pressureless sintering; the first major division is with respect to solid versus liquid phase processes.

The major distinction among pressureless sintering techniques is between solid state and liquid phase processes. Figure 1.1 provides a schematic definition of the various sintering techniques. Single-phase, solid state sintering has received the greatest consideration from a theoretical standpoint. Several reviews (1-5) give details on solid state sintering theory. Among the solid state processes there are several options involving second (solid) phases. These include compact homogenization (such as occurs with sintering mixed powders), activated sintering, and mixed phase sintering in the solid state. Activated sintering is a solid state analog to liquid phase sintering where a second solid phase contributes to rapid interparticle bonding (6). Mixed phase sintering occurs in an equilibrium two-phase field, such as with high carbon steels at temperatures where ferrite and cementite coexist. Homogenization occurs during sintering of mixed powders which form a single phase product, such as with alumina and chromia which form a total solid solution.

Although solid state sintering is the best understood form, liquid phase sintering has the greater industrial utilization. However, in spite of large-scale application, it is not well understood. The main treatment of liquid phase sintering by Eremenko et al. (7) was translated from a 1968 Russian publication. Thus, the documentation of liquid phase sintering has greatly lagged behind the industrial applications.

For this presentation, we will define liquid phase sintering as sintering involving a coexisting liquid and particulate solid during some part of the thermal cycle. There are two basic ways to obtain the liquid phase. The use of mixed powders of differing chemistries is the most common technique. The interaction of the two powders leads to the formation of a liquid during sintering. The liquid can result from melting of one component or formation of a eutectic. Furthermore, the liquid may be transient or persistent during sintering depending on the solubility relationship. Alternatively, a prealloyed powder can be heated to a temperature between the liquidus and solidus temperatures. The resulting mixture of liquid and solid phases leads to supersolidus sintering. Beyond these common forms of liquid phase sintering, there are some speciality techniques which will be treated in Chapter 7.

Within the main classes of liquid phase sintering, there are several possible variants dependent on the material characteristics. For example, the solid may be soluble or insoluble in the liquid. Such a difference greatly affects the rate of sintering and the microstructure evolution. Other major factors relate to the interfacial energies between the liquid and solid phases (wetting versus nonwetting liquids) and the relative penetration of the liquid along solid-solid grain boundaries. These variants coupled to the processing options like particle size, sintering temperature, time, atmosphere, and green density have further large effects on the type of material formed by liquid phase sintering. The cross-interaction of all of these factors contributes to the difficulty in studying liquid phase sintering. This variety also makes for a highly flexible manufacturing technique with widespread application to both metals and ceramics.

B. The Uses of Liquid Phase Sintering

As will be evident in this book, liquid phase sintering is used in several industrial and commercial products. In recent years there has been better understanding of the fundamental phenomena with concomitant increases in applications for liquid phase sintering. The earliest uses were in forming building bricks from clay based minerals (hydrated aluminum silicates) where a glassy phase formed the liquid. It is estimated that fired bricks were made up to 70 centuries ago (8,9). Subsequently, many other ceramic materials have been processed by liquid phase sintering including porcelain, earthenware, china, insulators, and refractories. These are typically characterized by the presence of a glassy phase at the sintering temperature. Furthermore, high resolution electron microscopy has documented that many of the modern technical ceramics also have glassy liquid phases at the grain boundaries during sintering. Today, a majority of ceramic products are fabricated with a liquid phase present during sintering, including abrasives, ferroelectric capacitors, ferrite magnets, electrical substrates, and high temperature covalent ceramics (10,11).

One of the first uses of liquid phase sintering for metals is attributed to the ancient Incas who converted platinum grains into a consolidated form by use of gold bonds. It is thought that the gold was molten during the sintering cycle. Artifacts from this process indicate its use over 400 years ago (12).

The development of modern liquid phase sintering technology is traced to the production of cemented carbides. Considerable effort went into the development of tool and machining materials in the 1900 to 1930 time period (13). By the early 1920's carbides with metallic binder alloys were patented. For these compositions sintering is with a liquid phase typically formed from iron, nickel, or cobalt. The liquid phase sintering approach permits the formation of dense, pore-free carbides with properties superior to any previously known cutting materials. Today the cemented carbides are an integral component of industrial operations including mining, machining, metal forming, grinding, drilling, and cutting. Their widespread use results from the composite properties of high strength, high hardness, low thermal expansion coefficient, and reasonable toughness.

Also during the 1920's bronze bearing materials were developed based on sintering mixtures of copper and tin powders. The classic oilless bronze bearings have an interconnected pore network created by the transient liquid phase when tin melts (14). These pores are subsequently filled with oil to provide self-lubrication during operation. Such bearings are most useful in low power electric motors and appliances.

The development of tungsten heavy alloys in the 1930's provided a theoretical basis for liquid phase sintering (15). These alloys were formed from mixtures of tungsten, nickel, and copper powders. Studies on heavy alloys provided a clear picture of the criteria for liquid phase sintering. Additionally, these alloys further demonstrated the unique property combinations in composite materials formed through liquid phase sintering. The use for heavy alloys today results from the combination of a high melting temperature, high strength, high density, ductility, corrosion resistance, and low thermal expansion coefficient. Such properties are useful for radiation shields, weights, inertial materials, machining supports, projectiles, and metalworking tools.

With the base provided by these earlier developments, recent applications of liquid phase sintering have expanded rapidly. The current uses include electrical contacts, wearfacing alloys, tool steels, superalloys, diamond-metal composites, dielectric materials, technical refractories, dental porcelain, magnetic materials, automotive structural components, aerospace components, and high temperature ceramics. To illustrate the varied interest in liquid phase sintering, Table 1.1 lists some examples of liquid phase sintered systems and their applications. These uses exist because of specific material benefits associated with the resulting microstructures and properties. From this industrial base, we can project a continued growth in the application of liquid phase sintering, especially in light of the improved understanding resulting from the current research and development activities.

C. Overview of Advantages and Limitations

From a technical point of view, the major advantage of liquid phase sintering is the result of faster sintering. The liquid phase provides for faster atomic diffusion than the concurrent solid state processes. The capillary attraction due to a wetting liquid gives rapid compact densification without the need for an external pressure. The liquid also reduces the interparticle friction, thereby aiding rapid rearrangement of the solid particles. In addition, liquid dissolution of sharp particle edges and corners allows more efficient packing. Grain size control is possible during liquid phase sintering; thus, processing effects can be carried over into microstructure manipulations to optimize properties. Finally, in many liquid phase sintering systems the higher melting phase is also the harder phase. This often results in sintered two-phase composite materials with ductile behavior in spite of a large quantity of hard phase.

However, there are some disadvantages. A common problem is compact slumping (shape distortion) which occurs when too much liquid is formed during sintering. Also, the same parameters which control the sintered microstructure often control the final properties. Separation of these effects is sometimes difficult. Furthermore, in the initial compact there are at least three phases; vapor, liquid, and solid. Accordingly, there are several interfaces and energies associated with such a structure. The several solubility, viscosity, and diffusivity effects coupled with the multiple phases have

TABLE 1.1

Examples of Liquid Phase Systems and Applications

System	Application
WC+Co	cutting and machining tools
Cu+Sn	oil-less bearings
Al+Pb	wear and bearing surfaces
W+Ni+Fe	radiation shields, weights
$Al_2O_3+SiO_2$	refractories for steelmaking
W+Ag	electrical contacts
Fe+Cu+C	structural components and gears
Ag+Hg	dental amalgam for fillings
Pb+Sn	soldering pastes
Fe+P	soft magnetic components
Al+Si+Cu	lightweight structural components
$BaTiO_3$+LiF	electrical capacitors
$Si_3N_4+Y_2O_3$	high temperature turbines

hindered a full analytic treatment of liquid phase sintering. Hot stage scanning electron microscopy demonstrates that the events associated with liquid phase sintering are very rapid (16,17). The rapid rates of sintering result in a predominantly qualitative treatment with less predictability than is found in solid state sintering.

In this book on liquid phase sintering, the focus will be on outlining the individual factors. A quantitative treatment is given where possible. But before developing a quantitative treatment, a qualitative description of classical liquid phase sintering is in order. This is followed by a description of the key microstructural features associated with sintered materials. Then the underlying thermodynamic and kinetic fundamentals are developed before detailing the various stages, forms, and applications of liquid phase sintering.

D. Classic Sequence of Stages

This section provides a qualitative description of the liquid phase effects on the sintered microstructure. In the usual case, the liquid wets the solid. Furthermore the liquid has a solubility for the solid. Consequently, the wetting liquid acts on the solid particles to eliminate porosity and reduce interfacial energy. The classic example of liquid phase sintering is attributed to Price et al. (15) in their report on tungsten heavy alloys. In this case the liquid persists throughout the high temperature portion of the sintering cycle, giving rapid densification and grain growth.

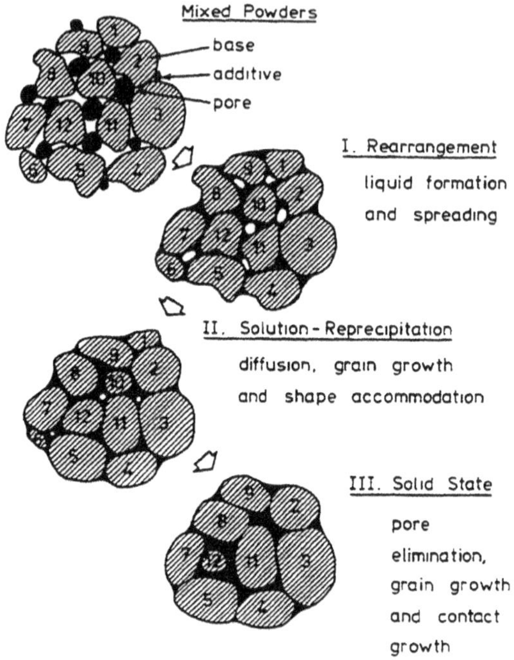

Figure 1.2 The classic stages of liquid phase sintering involving mixed powders which form a liquid on heating.

The classic liquid phase sintering system densifies in three overlapping stages (18-20). Figure 1.2 shows a schematic sequence of steps. Initially, the mixed powders are heated to a temperature where a liquid forms. With liquid formation there is rapid initial densification due to the capillary force exerted by the wetting liquid on the solid particles. The elimination of porosity occurs as the system minimizes its surface energy. During rearrangement, the compact responds as a viscous solid to the capillary action. The elimination of porosity increases the compact viscosity. As a consequence the densification rate continuously decreases. The amount of densification attained by rearrangement is dependent on the amount of liquid, particle size, and solubility of the solid in the liquid. Usually finer particles give better rearrangement. Full density (zero porosity) is possible by rearrangement if enough liquid is formed. It is estimated that 35 volume percent liquid is needed to obtain full density by rearrangement processes. However, rearrangement processes can be inhibited by a high green density or irregular particle shape. The particle contacts resulting from compaction form solid state bonds during heating, thereby preventing rearrangement.

Concurrent with the rearrangement stage are various other events. However, the kinetics of rearrangement are initially so fast that these other events are overshadowed. As densification by rearrangement slows, solubility

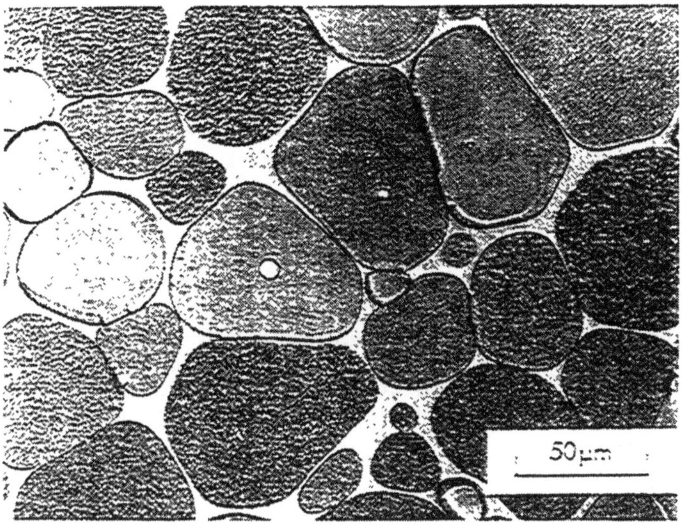

Figure 1.3 An optical micrograph of liquid phase sintered tungsten heavy alloy showing grain shape accommodation. The material is a 95% W-3.5% Ni-1.5% Fe alloy sintered at 1470°C for 2 h in hydrogen.

and diffusivity effects become dominant. This second stage of classic liquid phase sintering is termed solution-reprecipitation. A general attribute of solution-reprecipitation processes is microstructural coarsening. The coarsening is due to a distribution in grain sizes. The solubility of a grain in its surrounding liquid varies inversely with the grain size; small grains have a higher solubility than coarse grains. The difference in solubilities establishes a concentration gradient in the liquid. Material is transported from the small grains to the large grains by diffusion. The process is termed coarsening or Ostwald ripening. The net result is a progressive growth of the larger grains, giving fewer grains with a wider spacing. Solution-reprecipitation not only contributes to grain coarsening, but also to densification.

The grain shape can be altered by diffusion to allow tighter packing of the grains. This process of grain shape accommodation leads to pore elimination. The amount of liquid effects solution-reprecipitation in terms of both the diffusion distance and amount of grain shape accommodation. Both solubility of the solid in the liquid and diffusive transport are necessary criteria. The actual grain shape is determined by the relative solid-liquid and solid-solid interfacial energies, amount of liquid, and any anisotropy in surface energy of the solid. Figure 1.3 shows a typical liquid phase sintered microstructure with grain shape accommodation. The material is a W-Ni-Fe heavy alloy sintered to produce nearly pure tungsten grains in a matrix alloy of tungsten, nickel, and iron. In this case the tungsten grains are several times larger than the original particle size and have become flattened along neighboring faces. Although the solid-liquid interface area is enlarged by a deviation from a spherical shape, grain shape accommodation results in elimination of porosity and the higher energy interfaces associated with pores.

The last stage of classic liquid phase sintering is referred to as solid state controlled sintering. Densification is slow in this stage because of the existance of a solid skeleton. Processes dominant in the final stage are also active throughout the entire liquid phase sintering cycle; however, because of the slow nature, solid state sintering is not of significance until late in the sintering cycle. The rigidity of the solid skeleton inhibits further rearrangement, although microstructural coarsening continues by diffusion. The residual pores will enlarge if they contain an entrapped gas, giving compact swelling. During pore growth the pressure in the pores decreases. Besides these pore changes, the grain contacts allow solid state sintering. The diffusion events leading to contact growth between solid grains can be by solution-reprecipitation, coalescence of grains, or solid state diffusion. In general, properties of most liquid phase sintering materials are degraded by prolonged final stage sintering. Hence short sintering times are typically preferred in practice.

These three stages of classic (persistent) liquid phase sintering are summarized in Figure 1.4 in terms of the key microstructural changes. There are many possible variants of these stages and events. Such variants will be considered individually later in this book. However, the main concept of liquid phase sintering is evident, and provides the basis for focusing on the microstructure and the liquid-solid interactions discussed in the following chapters.

E. Nomenclature

A few terms are important to understanding observations made during liquid phase sintering. A brief review of these terms here will avoid later confusion. Density ρ is the mass per unit volume, and is often expressed as a fraction or percentage of theoretical density. Alternatively, the porosity ε is the fractional void space in the compact. By definition the fractional density plus the fractional porosity must equal unity,

$$\varepsilon + \rho = 1. \tag{1.1}$$

The green density refers to the pressed but unsintered condition of a powder compact. Thus, the green porosity corresponds to the initial void space prior to sintering. Theoretical density corresponds to a pore-free solid density. For a mixture of two phases, the theoretical density ρ_t can be estimated from the constituent densities as follows:

$$\rho_t = \rho_1 \rho_2/(X_1 \rho_2 + X_2 \rho_1) \tag{1.2}$$

where X is the weight fraction of each constituent and ρ_1 and ρ_2 are the constituent theoretical densities.

Densification is a useful concept in dealing with liquid phase sintered systems, especially when comparing systems of differing theoretical densities or initial porosities. Densification ψ is the change in porosity from the green

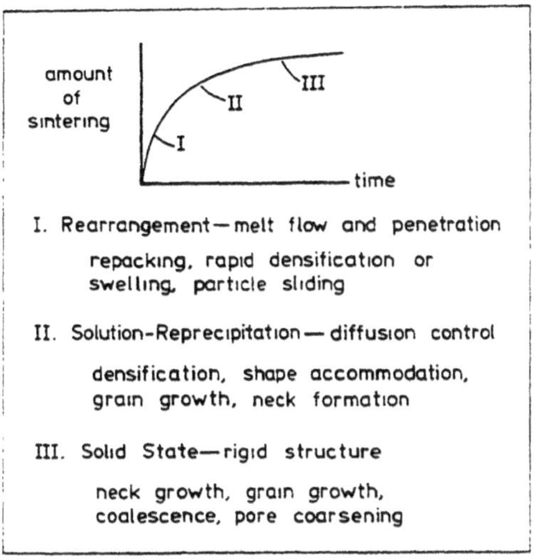

Figure 1.4 The process stages associated with classic liquid phase sintering, giving the main microstructural changes.

condition due to sintering, divided by the initial porosity,

$$\psi = (\varepsilon_g - \varepsilon_s)/\varepsilon_g \tag{1.3}$$

with the subscripts g and s representing the green and sintered conditions, respectively. Because densification is dimensionless, it is often expressed as a percentage. A densification of 100% corresponds to a compact which has been sintered to theoretical density.

Shrinkage refers to a decrease in linear dimensions, while swelling refers to an increase in dimensions. Two frequently applied measures of liquid phase sintering are the normalized linear and volumetric dimensional changes, $\Delta L/L_o$ and $\Delta V/V_o$, which are normalized by the initial length or volume.

The sintered density relates to the green density through the linear shrinkage as follows:

$$\rho_s = \rho_g/(1 - \Delta L/L_o)^3. \tag{1.4}$$

Measures such as density and shrinkage are easy to perform, and provide substantial insight into the rates of microscopic change during sintering. Variations in density in a powder compact are common. Compaction in rigid dies leads to nonisotropic green densities, with gradients in both the axial and radial directions (21). Thus, shrinkage or expansion can depend on the orientation of the measurement with respect to the compaction direction. For this treatment such variations in green density and dimensional change during sintering will be ignored.

Finally, in speaking about a powder mixture, the standard nomenclature will be to list the major phase first. Thus, WC-Co implies the bulk of the material is composed of tungsten carbide, with cobalt being the minor phase. Typically, the minor phase is responsible for forming the liquid and controls the amount of liquid formed during sintering.

F. References

1. H. E. Exner, "Principles of Single Phase Sintering," *Rev. Powder Met. Phys. Ceram.*, 1979, vol.1, pp.7-251.
2. G. C. Kuczynski, "Physics and Chemistry of Sintering," *Advan. Colloid Interface Sci.*, 1972, vol.3, pp.275-330.
3. F. Thummler and W. Thomma, "The Sintering Process," *Metall. Rev.*, 1967, vol.12, pp.69-108.
4. M. B. Waldron and B. L. Daniell, *Sintering*, Heyden and Sons, London, UK, 1978.
5. M. F. Yan, "Sintering of Ceramics and Metals," *Advances in Powder Technology*, G. Y. Chin (ed.), American Society for Metals, Metals Park, OH, 1982, pp.99-133.
6. R. M. German and Z. A. Munir, "Activated Sintering of Refractory Metals by Transition Metal Additions," *Rev. Powder Met. Phys. Ceram.*, 1982, vol.2, pp.9-43.
7. V. N. Eremenko, Y. V. Naidich, and I. A. Lavrinenko, *Liquid Phase Sintering*, Consultants Bureau, New York, NY, 1970.
8. M. Chandler, *Ceramics in the Modern World*, Doubleday, Garden City, NY, 1968.
9. F. H. Norton, *Elements of Ceramics*, second edition, Addison-Wesley, Reading, MA, 1974.
10. D. W. Budworth, *An Introduction to Ceramic Science*, Pergamon Press, Oxford, UK, 1970.
11. W. D. Kingery, H. K. Bowen, and D. R. Uhlmann, *Introduction to Ceramics*, second edition, Wiley-Interscience, New York, NY, 1976.
12. P. Bergsoe, "The Metallurgy and Technology of Gold and Platinum Among the Pre-Columbian Indians," *Ingeniorvidenskabelige Skrifter*, 1937, vol.A44, pp. 1-44.
13. P. Schwartzkopf and R. Kieffer, *Cemented Carbides*, MacMillan, New York, NY, 1960.
14. V. T. Morgan, "Bearing Materials by Powder Metallurgy," *Powder Met.*, 1978, vol.21, pp.80-85.
15. G. H. S. Price, C. J. Smithells, and S. V. Williams, "Sintered Alloys. Part I - Copper-Nickel-Tungsten Alloys Sintered with a Liquid Phase Present," *J. Inst. Metals*, 1938, vol.62, pp.239-264.
16. L. Froschauer and R. M. Fulrath, "Direct Observation of Liquid-Phase Sintering in the System Tungsten Carbide-Cobalt," *J. Mater. Sci.*, 1976, vol.11, pp.142-149.
17. L. Froschauer and R. M. Fulrath, "Direct Observation of Liquid-Phase Sintering in the System Iron-Copper," *J. Mater. Sci.*, 1975, vol.10, pp.2146-2155.
18. F. V. Lenel, "Sintering in the Presence of a Liquid Phase," *Trans. AIME*, 1948, vol.175, pp.878-896.

19. H. S. Cannon and F. V. Lenel, "Some Observations on the Mechanism of Liquid Phase Sintering," *Plansee Proceeding*, F. Benesovsky (ed.), Metallwerk Plansee, Reutte, Austria, 1953, pp.106-121.
20. W. D. Kingery, "Densification During Sintering in the Presence of a Liquid Phase. 1. Theory," *J. Appl. Phys.*, 1959, vol.30, pp.301-306.
21. E. Mosca, *Powder Metallurgy Criteria for Design and Inspection*, Associozione Industriali Metallurgici Meccanici Affini, Turin, Italy, 1984.

CHAPTER TWO

Microstructures

A. Typical Microstructures

The variety of materials fabricated by liquid phase sintering results in a spectrum of sintered microstructures as exemplified in Figure 2.1. These micrographs include a cemented carbide, porous bronze, and structural iron alloy. There are obvious differences in the shape and distribution of phases. The goal of this chapter is to define the microstructural characteristics of materials processed by liquid phase sintering. Furthermore, quantitative links between several of the microstructural parameters are established in this chapter.

There are several techniques available to monitor development of materials processed by liquid phase sintering. These include microstructure, density, electrical conductivity, magnetic behavior, hardness, strength, toughness, elastic modulus, and x-ray analysis. In spite of this array of techniques, microstructural observations are the most significant. From microstructural analysis we can measure the amount of each phase, size of the solid grains, solid-solid contact size, porosity, and grain shape. Microstructural measurements also provide insight into the basic thermodynamic behavior of the materials. Measurements made as a function of sintering time or sintering temperature allow monitoring of the kinetics. Other observations, such as those of strength or toughness, are most useful when combined with microstructural measurements.

In multiple phase materials, microstructure is characterized by grain shape, grain size, grain orientation, relative amount of each phase, and interactions between the phases. In liquid phase sintered materials there is a distribution in all of these properties. For example, in order to minimize interfacial energies, changes in grain size and shape are common at elevated temperatures. In some cases this results in deviations from a simple spherical grain shape. Furthermore, the resulting phase distribution is dependent on relative interfacial energies. Figure 2.2 contrasts the microstructures of sintered copper with 20 volume percent of either lead or bismuth. The copper-bismuth system has a lower solid-liquid interfacial energy; thus, bismuth is distributed throughout the microstructure. The lead, on the other hand, has a higher solid-liquid energy which inhibits penetration between the copper particles. As can be seen, some of the lead remains segregated as islands in the sintered microstructure. Beyond the interfacial energy effect, the alloy composition has an influence on the sintered microstructure. This is

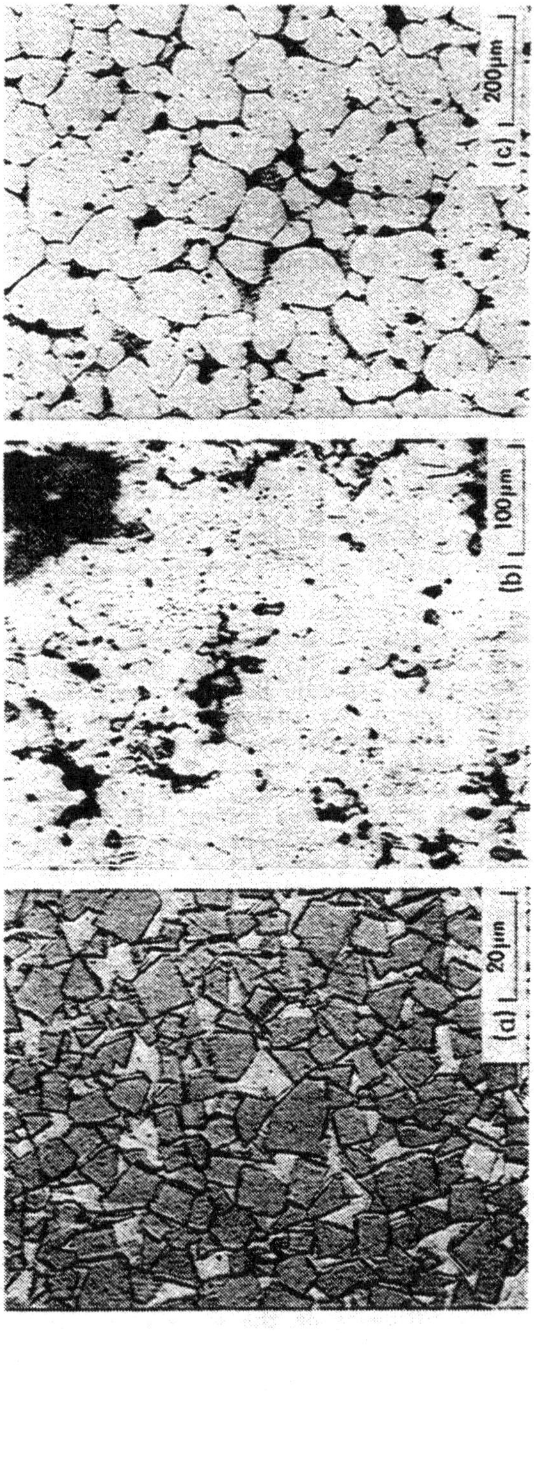

Figure 2.1 Optical micrographs of materials processed by liquid phase sintering; (a) WC-Co, (b) Cu-Sn, and (c) Fe-Ni-Mo-C-B alloy (photos courtesy of J. Strauss, N. Gendron, D. Madan).

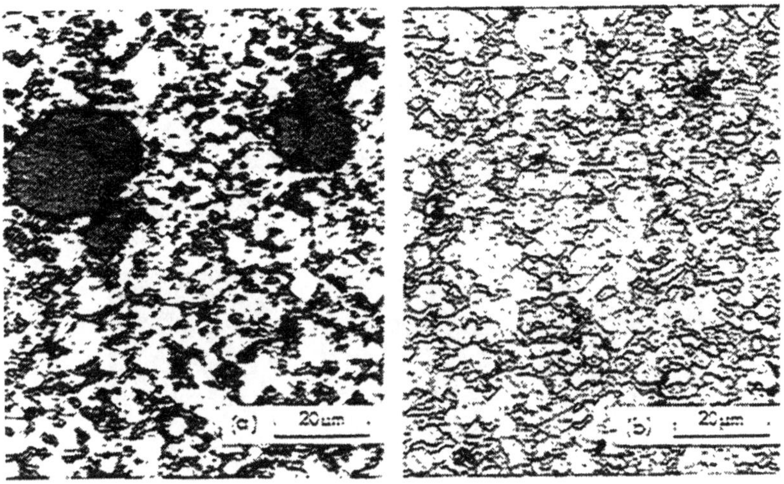

Figure 2.2 Optical micrographs contrasting the microstructural differences between liquid phase sintered Cu-Pb (a) and Cu-Bi (b) with 20% liquid phase.

illustrated by the cobalt-copper alloys in Figure 2.3. These drawings are based on studies (1,2) covering the range from 30 to 80 wt.% cobalt. In this system, cobalt remains as the solid while copper forms a wetting liquid at temperatures above 1083°C. At low cobalt levels, the cobalt grains are small, separated by the solidified matrix, and are nearly spherical in shape. As the cobalt concentration increases, the grain size becomes larger, grain contacts become more frequent, and at high cobalt concentrations the grains become less spherical. The quantitative description of these several microstructural variations with composition and interfacial energy are the focus of this chapter.

B. Contact Angle

When a liquid forms during liquid phase sintering the microstructure consists of at least three phases - solid, liquid, and vapor. Wetting describes the equilibrium between the three phases. The contact angle is a physical characteristic which represents a balance between the interfacial energies. For a liquid to wet a solid, the total free energy must be decreased. Good wetting corresponds to the case shown in Figure 2.4a, while poor wetting is shown in Figure 2.4b. The degree of wetting is characterized by the contact angle shown in each of these figures. The contact angle θ is the included angle in the liquid; its magnitude depends on the balance of surface energies. As shown in Figure 2.4c, the contact angle is associated with the vector equilibrium of the three interfacial energies. For the point of solid-liquid-vapor contact to be in equilibrium, the horizontal components of the three vectors must equal zero. Thus,

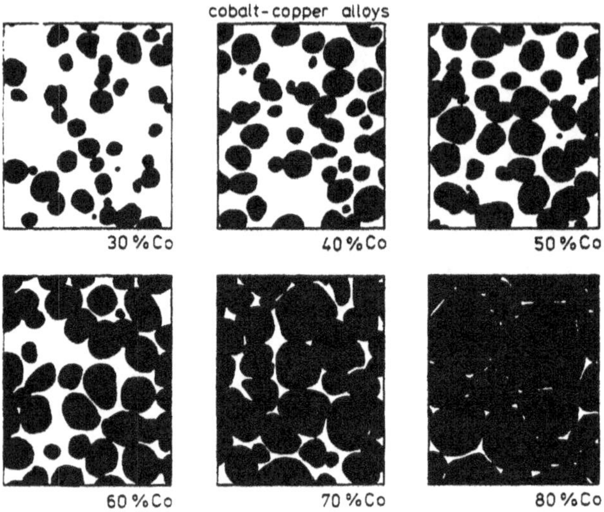

Figure 2.3 Microstructures of liquid phase sintered Co-Cu alloys with various concentrations of cobalt (the dark phase) based on the studies of Kang and Yoon (1,2).

Figure 2.4 The solid-liquid-vapor equilibrium for good wetting (a) and poor wetting (b) situations. The relation between the contact angle and the three interfacial energies is given in (c).

$$\gamma_{SV} = \gamma_{SL} + \gamma_{LV} \cos\Theta \tag{2.1}$$

or in terms of the contact angle,

$$\Theta = \arccos[(\gamma_{SV} - \gamma_{SL})/\gamma_{LV}]. \tag{2.2}$$

The subscripts *S*, *L*, and *V* represent solid, liquid, and vapor, respectively.

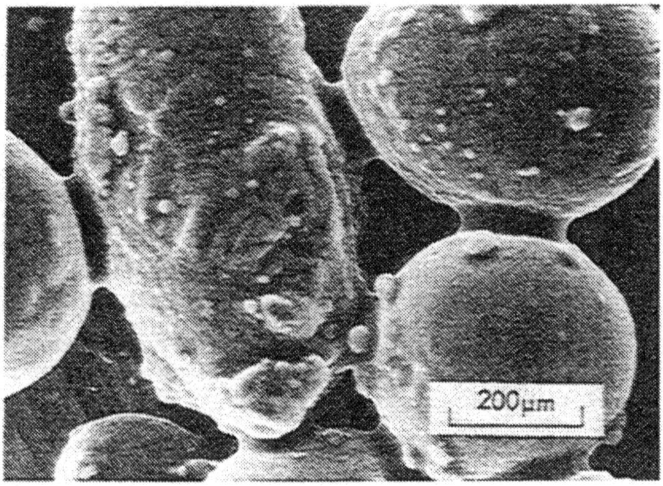

Figure 2.5 A scanning electron micrograph showing the bridging between solid grains associated with a wetting liquid during the initial stage of liquid phase sintering.

Note that the contact angle depends on the difference in interfacial energies and not on the absolute values. Thus, knowledge of a high or low surface energy provides no insight into wetting behavior. For many materials, the contact angle and the interfacial energies are dependent on surface purity. Surface impurities are common with commercial powders; hence, wetting behavior during liquid phase sintering can be drastically altered by contaminants or processing steps which clean the powder surface.

Wetting is associated with a chemical reaction at the interface. Wetting is aided by solubility of the solid in the liquid, formation of intermediate compounds, and interdiffusion. For powders, a typical solid-liquid-pore configuration can be seen in Figure 2.5. This scanning electron micrograph shows the wetting liquid bridging between the solid particles. Such a wetting liquid provides substantial bonding force on the solid particles. In typical powder systems there is a range of particle sizes, pore sizes, pore shapes, and particle shapes which results in a range of capillary conditions. A wetting liquid will attempt to occupy the lowest free energy position; thus, it preferentially flows to the smaller capillaries which have the highest energy per unit volume. When there is insufficient liquid to fill all of the pores, the wetting liquid will attempt to pull the particles together to minimize the free energy. This effect gives rise to the rearrangement stage and rapid initial densification in liquid phase sintering.

In contrast, poor wetting means that the presence of the liquid on the solid surface is unfavorable. In liquid phase sintering the formation of a poor wetting liquid may lead to swelling of the compact during heating, and possibly melt exuding from surface pores. Thus, depending on the contact angle, liquid formation during sintering can cause either densification or swelling. The overall magnitude of the capillary effect depends on the amount of liquid, size of particles (and pores), contact angle, and particle shape.

C. Dihedral Angle

The dihedral angle ϕ is formed where a solid-solid grain boundary intersects the liquid. The dihedral angle is important to the microstructure of polycrystalline grains and to grain-grain contacts in the liquid phase. There is no dihedral angle for an amorphous solid.

Figure 2.6a shows a general three phase junction, where the interfacial energies are denoted by vectors. There are three interfacial energies and three angles forming an equilibrium balance. As shown in Figure 2.6b, the three angles and three energies form a triangle. Applying the law of sines gives

$$\gamma_{12}/\sin\phi_3 = \gamma_{23}/\sin\phi_1 = \gamma_{13}/\sin\phi_2 \tag{2.3}$$

where the general substitution of

$$\sin(180° - x) = \sin(x) \tag{2.4}$$

has been made for each angle. Equation (2.3) is known as the Dupre equation.

In the case typical to liquid phase sintering, two of the energies are equal. This situation is shown in Figure 2.7, with the solid-solid grain boundary energy shown in opposition to the two solid-liquid interface energies. Applying Equation (2.3) to the condition established in Figure 2.7 gives,

$$\gamma_{SL}/\sin(\phi/2) = \gamma_{SS}/\sin\phi. \tag{2.5}$$

Applying the trigonometric identity of sin(2x) = 2 sin(x) cos(x) gives,

$$2\ \gamma_{SL}\ \cos(\phi/2) = \gamma_{SS} \tag{2.6}$$

which is the Young equation, linking the dihedral angle to the two interfacial energies. Rearranging Equation (2.6) gives the dihedral angle as

$$\phi = 2\ \arccos[\gamma_{SS}/(2\ \gamma_{SL})]. \tag{2.7}$$

Thus, the dihedral angle is characteristic of the energy ratio between the grain boundaries and solid-liquid surfaces. An alternative derivation of Equation (2.6) can be obtained by summing the vertical vector components in Figure 2.7. At equilibrium, the sum will equal zero.

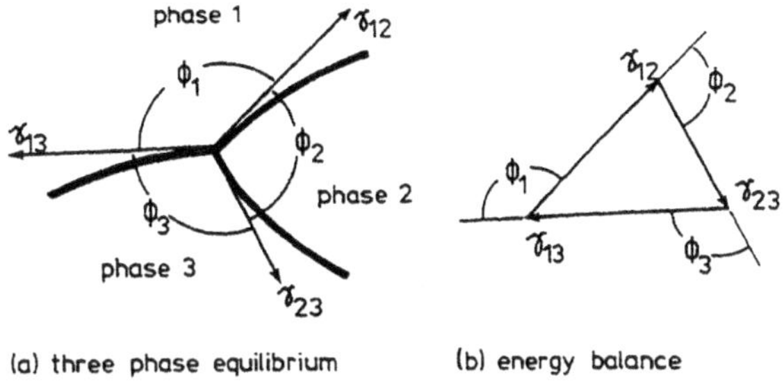

Figure 2.6 The general equilibrium at a three-phase intersection. The interfacial energies and angles are shown in (a), while the equilibrium energy triangle is shown in (b).

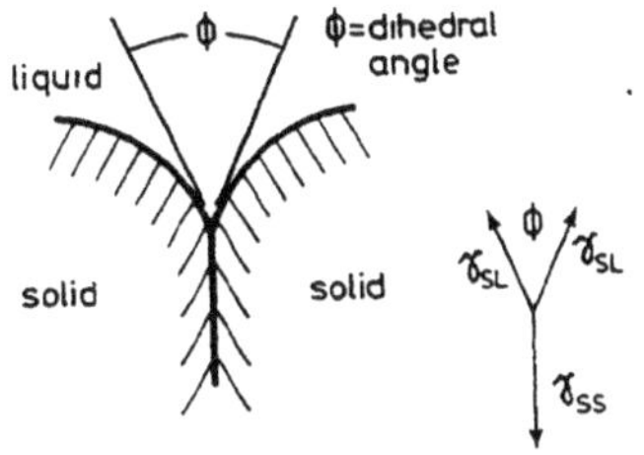

Figure 2.7 The dihedral angle and surface energy equilibrium between two intersecting grains with a partially penetrating liquid phase.

For large-angle grain boundaries, the dihedral angle is a solid-liquid characteristic. Figure 2.8 shows some examples of three grain systems with a liquid at the triple point. The dihedral angle affects both the liquid and grain shapes. Note that as the dihedral angle approaches 0°, the solid-solid to solid-liquid energy ratio approaches 2. Figure 2.9 shows the dihedral angle variation with this energy ratio from 0 to 2. If this energy ratio is greater than 2, the dihedral angle is 0°, as shown in Figure 2.9, and the liquid will penetrate the grain boundaries of the solid phase. Thus, no static equilibrium exists involving grain boundaries in the presence of the liquid. The other extreme occurs when the energy ratio is small; then no liquid penetration of the grain boundaries is possible.

So far we have considered the interfacial energies to be essentially constant. Measurements on crystalline solids show that the solid-liquid surface energy varies with crystallographic orientation. Usually this effect gives a total variation of less than 20% in the solid-liquid surface energy over all possible orientations. (The effect of anisotropic surface energy on grain shape is discussed later in this chapter.) More significant are the variations in the solid-solid grain boundary energy with crystallographic orientation. It is possible for solid-solid contacts to form low-angle grain boundaries. In

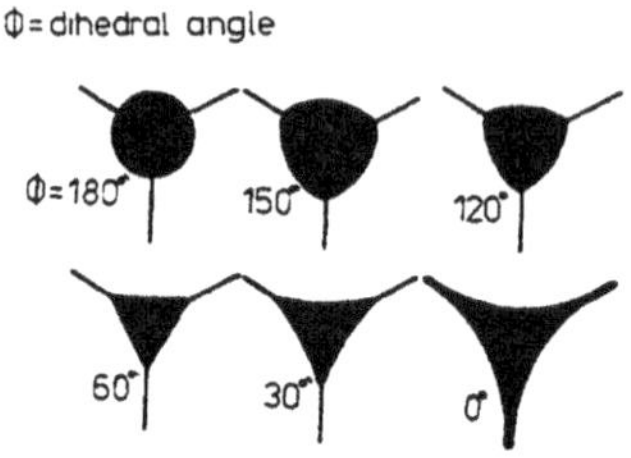

Figure 2.8 The effect of the dihedral angle on the liquid shape at the intersection of three solid grains.

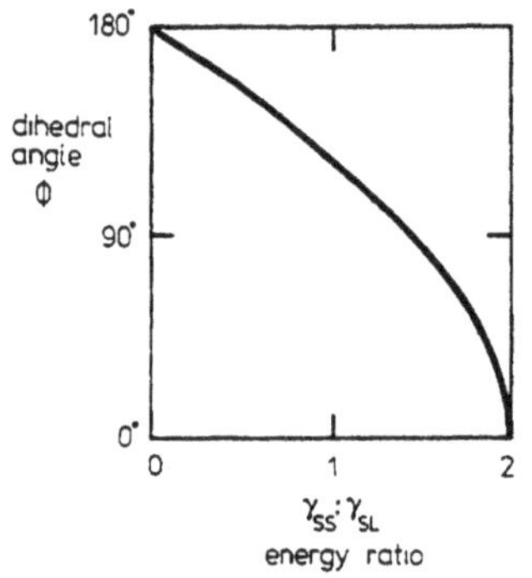

Figure 2.9 The dihedral angle variation with the solid-solid to solid-liquid energy ratio. For an energy ratio over 2, the dihedral angle will be 0° with penetration of grain boundaries.

such cases the grain boundary energy varies with the misorientation up to approximately a 10° misorientation. If α is the misorientation angle between contacting grains, then

$$\gamma_{SS} = \alpha[A - B \ln(\alpha)] \tag{2.8}$$

where A and B are constants relating to the dislocation energy (3). According to Equations (2.7) and (2.8), for small misorientation angles the dihedral angle will be quite large and will tend toward 180°. The low-angle grain boundaries often form by particle-particle contact during liquid phase sintering. This contact leads to possible grain coalescence, giving larger grain sizes as time progresses.

The dihedral angle cannot be measured directly on two-dimensional micrographs. The random orientation of the cross section with respect to the grain contacts cannot assure observation of the true angle. However, Riegger and Van Vlack (4) show that the mean dihedral angle as measured on two dimensional sections is the best estimate of the true value. Thus, measurements on several random cross sections provide an estimate of the true value with a typical error of 5°. A distribution in the dihedral angle is also possible. Such a distribution can result from anisotropic surface energies and various misorientation angles.

Densification in liquid phase sintering depends on both the dihedral angle and the contact angle. Generally, better liquid phase sintering is associated with smaller values of both angles.

So far the dihedral angle has been treated as a constant, dependent on the interfacial energies and misorientation. In reality it is also a function of time in liquid phase sintering. With melt formation and spreading, grain boundary penetration will lead to a chemistry change in the liquid. During solid dissolution up to the solubility limit, the liquid chemistry and interfacial energy will be changing. With time, the solid-liquid surface energy will stabilize at a value lower than the initial (unalloyed) melt value. Indeed, the dissolution reactions during liquid spreading decrease the solid-liquid interfacial energy below the equilibrium value. Such behavior is predicted by the model of Aksay et al. (5). After spreading and reaction have come to completion, the system approaches an equilibrium interfacial energy and an equilibrium dihedral angle. Figure 2.10 demonstrates the transients in the solid-liquid interfacial energy and the corresponding dihedral angle variations. Because of the rapid initial decrease in dihedral angle, there is a possible over-penetration of the liquid into grain boundaries. Such microstructural transients have been documented in systems like tungsten-nickel (6) and cemented carbides (7).

D. Volume Fraction

For a liquid phase sintered system, there are initially three phases - solid, liquid, and pore. Each of these phases will initially occupy a fraction of the total volume. Thus, there are at least three volume fraction parameters of concern, the volume fractions of solid, liquid, and pore, denoted as V_S, V_L, and V_P, respectively.

Generally the objective during liquid phase sintering is for the porosity to approach zero. For persistent liquid phase sintering the volume fractions of solid and liquid become constant at long sintering times. However, in reactive and transient liquid phase sintering all three volume fractions are changing with sintering time. The three volume fractions must sum to unity,

$$V_S + V_L + V_P = 1. \tag{2.9}$$

The microstructural changes associated with an increasing volume fraction of solid are quite pronounced. Figure 2.3 illustrates the grain shape and grain size changes with increasing volume fraction of solid. In addition, more solid-solid contacts form at the higher volume fractions of solid.

In analyzing the volume fraction of a phase, it is generally assumed that the structure is random. This means that there will be no sectioning effect. However, gravitational settling does occur if the solid and liquid differ in density (8). The most typical technique for measuring the amount of

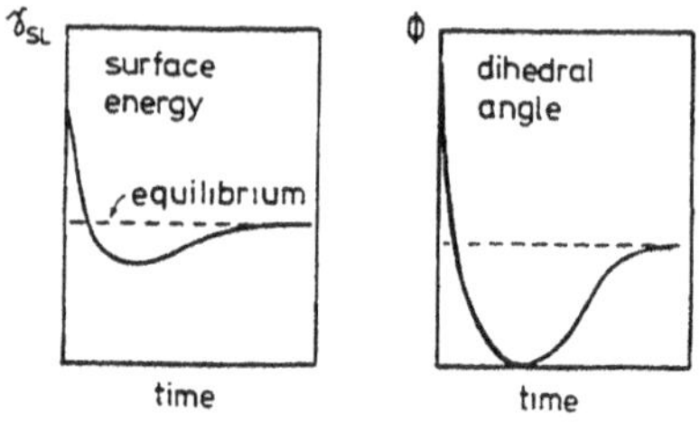

Figure 2.10 The variation in solid-liquid surface energy and dihedral angle with time. As the solid-liquid reaction progresses the two parameters will approach their equilibrium values.

each phase is to apply point count analysis to a polished cross section. A benefit of this approach is that no shape assumption is necessary. (A common difficulty with such an approach is distortion of the microstructure during polishing and etching.) Underwood (9) provides a review of the various analytical techniques available for determining the amount of a phase. If the solid phase is spherical, Fullman (10) shows a simple formula for determining volume fraction,

$$V_S = [8\ N_L^2/(3\ \pi\ N_A)] \tag{2.10}$$

where the number of grains per unit length and area are denoted as N_L and N_A, respectively.

E. Porosity and Pore Shape

Pores are an inherent part of liquid phase sintering. Pores are present in the powder compact as interparticle voids. Also, pores can result from uneven liquid distribution, unbalanced diffusion events (the Kirkendall effect), reactions with the vapor, and capillary spreading of the liquid upon melting. The porosity, or void space, is characterized by an amount, size, shape, and distribution throughout the compact.

On liquid formation there are pores in the initial compact. As the liquid flows and spreads into narrow capillaries there may be adequate compact shrinkage to eliminate pores. However, in some cases spreading appears not to be uniform (11). Rather, the liquid tends to occupy the compact center and spread outward, giving a radial increase in porosity. For a wetting liquid, the smallest pores are filled first because of the favorable interfacial energy reduction (unless the liquid has a high contact angle). The preferential filling of small pores at the expense of large pores can cause the formation of large pores at sites occupied by the melt-forming particles (12). Figure 2.11 illustrates the consequence of this event. The liquid flow has been radially outward, leaving behind a spherical pore. In some cases the melt spreading can actually cause swelling instead of the anticipated densification. Such swelling is minimized by making the melt-forming particles small, on the scale of the interparticle voids.

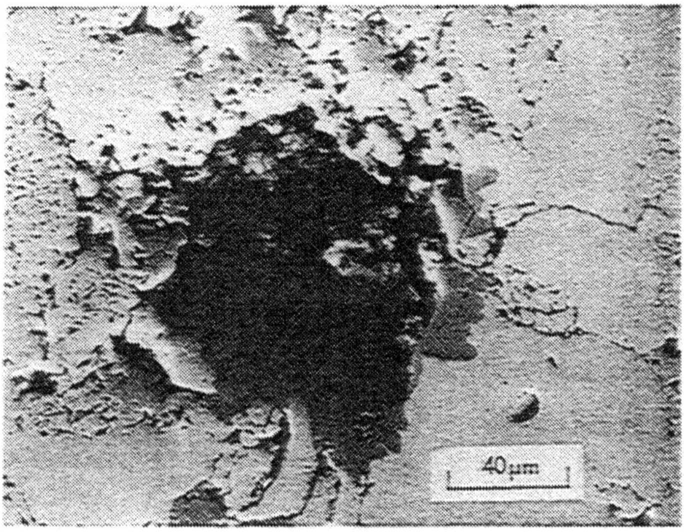

Figure 2.11 Pore formation at a prior additive particle site in a powder mixture of iron and aluminum due to capillary action after heating to 685°C (photo courtesy of D. J. Lee).

Later, the large pores can be filled by a process of meniscus growth (11-13). Densification to zero porosity is energetically favorable, but trapped gas in the pores can inhibit final densification (14). Also, the formation of a rigid solid skeleton in the final stage will hinder pore elimination.

Characterization of the pores is possible by several techniques. Metallography is usually most suitable. Mercury porosimetry can be applied to open pore structures (those pores connected to the compact surface are termed open pores). In addition, tests using impregnating fluids or epoxy resins allow three-dimensional visualization of the open pores after the other phases are etched away (15,16).

The shape of the pores varies rapidly during liquid phase sintering. In the first stage, the pores are irregular. Later they form a cylindrical network and finally attain a spherical shape. The pore network will decompose from complex interconnected cylindrical shapes as densification occurs. At roughly 8% porosity, the cylindrical pores are expected to pinch off into more stable spherical shapes (14).

F. Grain Size

The grain size, like the pore size, is a distributed parameter. In a two dimensional cross section, the apparent grain size is smaller than the true grain size. This is due to the randomness of the intersection of the grains, only a few of which are cut at their largest diameter. If the grains are uniform spheres, then the grain size can be determined from micrograph

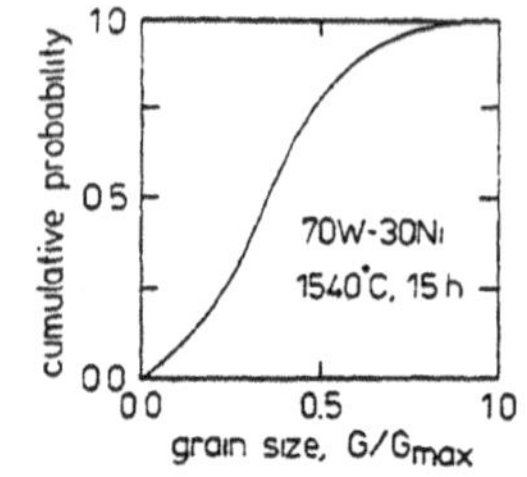

Figure 2.12 The normalized cumulative grain size distribution for W-Ni after prolonged sintering. The maximum grain size was 140 μm after 15 hours at 1540°C (17).

measurements. For spheres of equal size, the grain size G is given in terms of the number of grains intersected per unit length of test line N_L and the number of grains per unit cross sectional area N_A *(10)*,

$$G = (4/\pi)\ (N_L/N_A). \tag{2.11}$$

Instead of dealing with the grain size, it is usually more convenient to deal with the intercept size, defined in terms of the intercept length for a random test line passing over a two dimensional cross section. The mean intercept size L is defined as the ratio of the volume fraction of solid to the number of grain intercepts per unit length,

$$L = V_S/N_L. \tag{2.12}$$

For a single phase material, the mean grain size is inversely proportional to the number of grain intercepts per unit length of test line. Usually the mean intercept length is sufficient to describe the grain size. However, in some instances the coarser fraction of the grain size distribution is important (such as in fracture). In such cases it is necessary to determine the distribution by careful measurements. An example of a grain size distribution following liquid phase sintering is shown in Figure 2.12. The cumulative probability is shown versus the grain size on a normalized scale. The typical normalized scale divides the actual sizes by the maximum observed size. These data are based on the measurements of Kang and Yoon (17) for a tungsten-nickel alloy sintered at 1540°C. Distribution information like this can be used to isolate the mechanism of grain coarsening. One problem with such an analysis is that the theories assume a spherical shape with a wide separation; such a grain shape is often not obtained in liquid phase sintering.

Grain size can be expressed in any of several units. The easiest to understand is the mean intercept length. The mean grain diameter implies a spherical shape (or an assumed spherical shape). Other measures of grain size are grains per unit area and grains per unit volume.

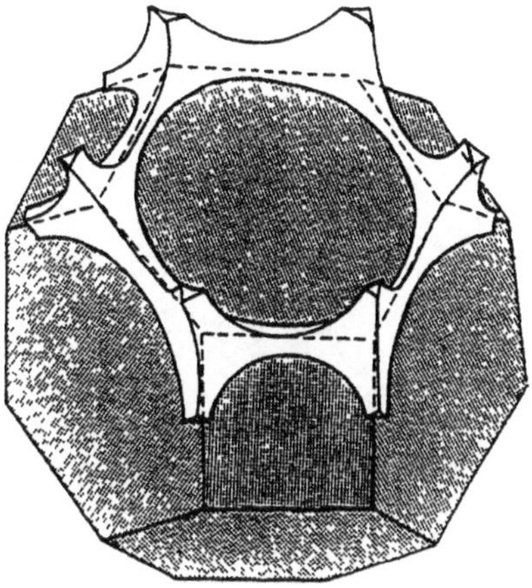

Figure 2.13 A drawing of a segment of a connected, partially penetrating liquid on grain edges where the grain is a tetrakaidecahedron (14 sides) (18).

G. Grain Shape

The grain, which is solid during liquid phase sintering, has a shape dependent on the volume fraction of solid, dihedral angle, and surface energy anisotropy. Contacts form between neighboring grains, especially at the higher volume fractions of solid. These contacts cause the grains to change shape to attain better packing. For dihedral angles over 60° and small volume fractions of liquid, the liquid structure is dispersed along grain edges and is not continuous (18-21). At large dihedral angles, typically over 90°, the microstructure is unstable for all quantities of liquid. As a consequence, the liquid will exude from the compact. More practical interest exists in systems with dihedral angles below 60°. There are minimum energy configurations involving the dihedral angle, volume fraction of liquid, and grain shape.

Beere (18) has solved for the grain-liquid equilibrium shape under various conditions assuming zero porosity and a packing coordination of 14. His results show that a dihedral angle over 30° requires a proportionate increase in the amount of liquid to attain full liquid linkage along grain edges. For a dihedral angle over 60° the liquid becomes isolated at the triple points between grains. An example of the grain-liquid shape according to the calculations by Beere is shown in Figure 2.13. The liquid forms a continuous network along the three grain junctions. Figure 2.14 is a scanning electron micrograph of such a network after chemical dissolution of the grain structure. The interlinked liquid (matrix) structure is evident in this micrograph.

Park and Yoon (19) have solved for the minimum energy configurations using an assumed grain packing coordination of 12. The results of their calculations are shown in Figures 2.15 and 2.16. The energy for a configuration versus the volume fraction of liquid is shown in Figure 2.15 for various

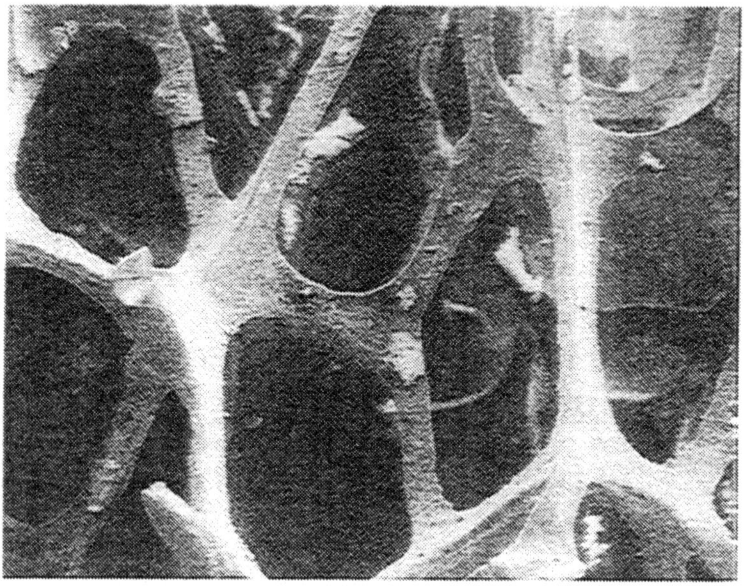

Figure 2.14 A scanning electron micrograph of the matrix network after liquid phase sintering. The solid grains have been removed by chemical dissolution (photo courtesy J. T. Strauss).

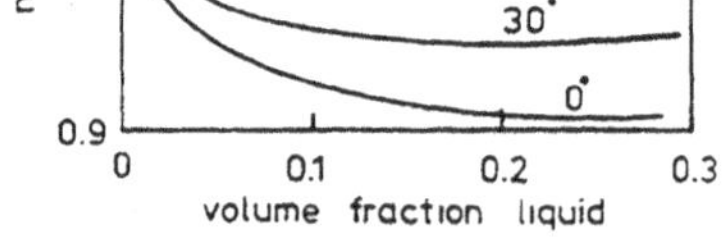

Figure 2.15 The variation in system energy for solid grains in a liquid matrix as calculated for various dihedral angles and volume fractions of liquid with a coordination of 12 (19).

dihedral angles. There is a minimum energy point on the low dihedral angle curves, which leads to the plot shown in Figure 2.16. Thus, there is a combination of dihedral angle and volume fraction of solid for minimum energy. Above a dihedral angle of approximately 60° no minimum energy configuration exists. In comparison to Beere (18), there are some differences in the minimum energy conditions due to the different calculation techniques and assumed grain packing geometries.

Wray (20) has considered the conditions leading to a connected microstructure for an assumed grain coordination of 14. Examples of six equilibrium liquid structures are shown in Figure 2.17 based on Wray's calculations. These structures correspond to the six regions on the dihedral angle-volume fraction liquid map also shown in Figure 2.17. A dihedral angle below 70.53° leads to a concave liquid. With a sufficient amount of liquid, the structure can be connected along the grain edges, and will be connected for all dihedral angles below 60° independent of the amount of liquid. Alternatively, for low volume fractions of liquid and large dihedral angles, an unconnected liquid microstructure is expected. The relation between connectivity, dihedral angle, and volume fraction of liquid is given by the contours shown in Figure 2.17. Note that these equilibrium microstructures assume isotropic surface energies and assume all porosity has been removed from the compact. The relations are independent of grain size.

As liquid phase sintering progresses, the large grains will grow at the expense of the smaller grains. Dissolution will tend to make the smaller grains spherical (22). This is because the rate of size change is often too rapid during dissolution for the grain to maintain an equilibrium shape. However, growing grains tend toward shapes dictated by the Gibbs-Wulff plot of surface energy versus crystallographic orientation (23). That is, low energy crystallographic orientations are favored, leading to facting of the grains.

The flattening of normally spherical grains because of a high volume fraction of solid is illustrated in Figure 2.3. At a low volume fraction of solid the grains are often rounded, approaching a spherical shape. While at high volume fractions of solid the grains show flattened faces. Such shape accommodation is necessary to eliminate porosity at low dihedral angles and large volume fractions of solid; Figure 2.18 shows how these two parameters interact to define the regions where shape accommodation is necessary. In no case can a low dihedral angle system be fully densified with a high volume fraction of solid without shape accommodation. The particle shape before liquid phase sintering has no significant effect on the sintered grain shape for traditional systems. Because of solution and reprecipitation during liquid phase sintering, the particles are rounded and change shape during growth.

These various results correspond to model grain packing involving uniform sized grains. In systems where the surface energy varies with crystallographic orientation by more than approximately 15%, irregular grain shapes are expected. This effect is most prevalent in the cemented carbides, such as shown in Figure 2.19. The prismatic grain shape is dictated by the anisotropic surface energy. A sketch of the anisotropic tungsten carbide grain shape is given in Figure 2.20. The observed grain shape in a random cross section will depend on the grain orientation with respect to the section plane. A variety of corresponding grain shapes are evident in the micrograph

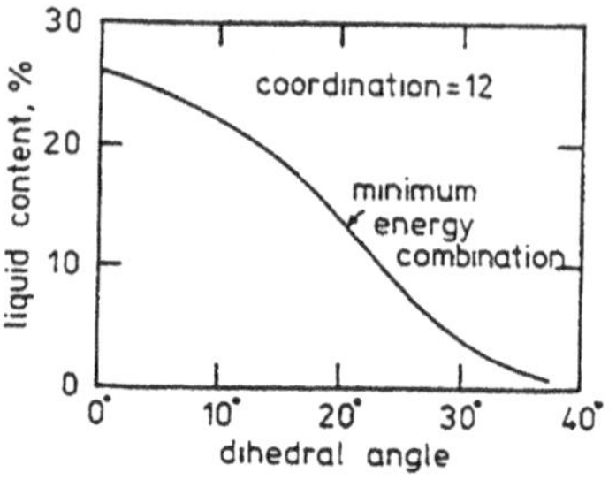

Figure 2.16 The minimum energy configuration from Figure 2.15 showing the volume fraction solid variation with dihedral angle for a packing coordination of 12.

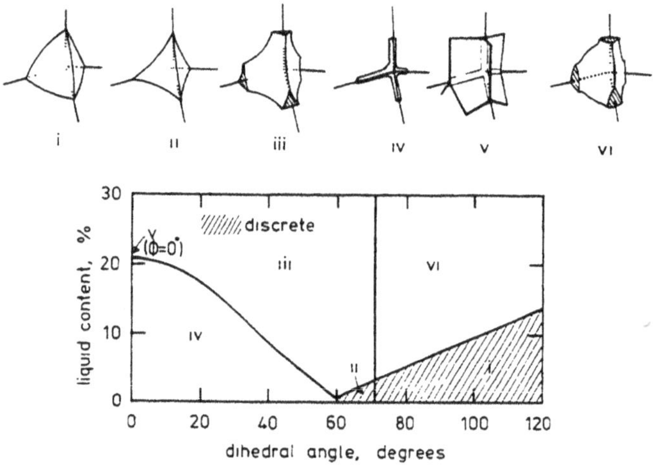

Figure 2.17 The calculated grain-liquid configurations of Wray (20) as dictated by the combinations of volume fraction and dihedral angle. The upper drawings correspond to the regions in the lower plot.

of a sintered WC-Co cemented carbide shown in Figure 2.19. Additionally, impurities and chemical additives can alter the grain shape through their effect on the surface energy.

Anisotropic surface energies lead to a nonspherical grain shape even for low volume fractions of solid. For such cases Warren (24) provides a example calculation of grain shape assuming the surface energy of the (100) plane deviated from a mean value for the crystal. Figure 2.21 gives the resulting grain shape as a function the surface energy variation from the mean. A change from an angular to a rounded grain shape occurs over a small range

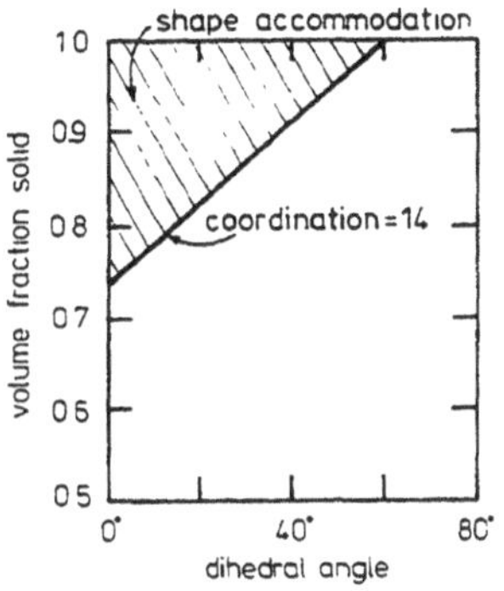

Figure 2.18 The relation between dihedral angle and volume fraction of solid, showing the region where grain shape accommodation is necessary, assuming a packing coordination of 14.

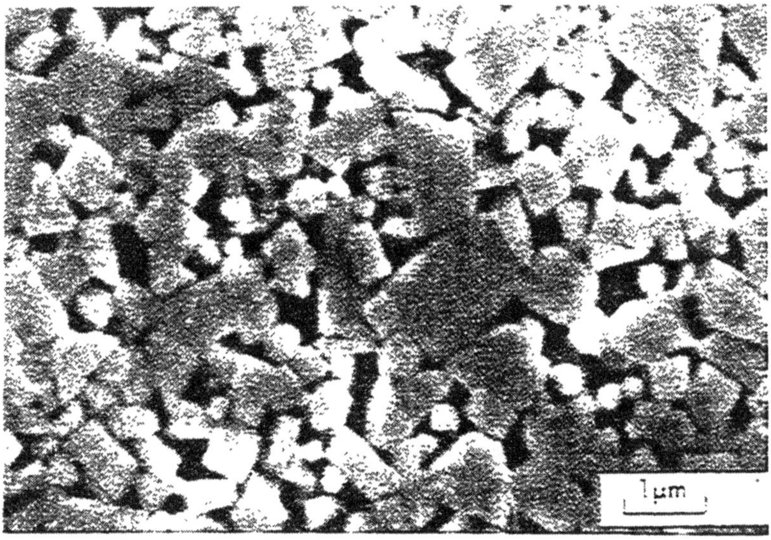

Figure 2.19 A scanning electron micrograph of a sintered cemented carbide with prismatic WC grains in a matrix based on Co (photo courtesy of G. Dixon).

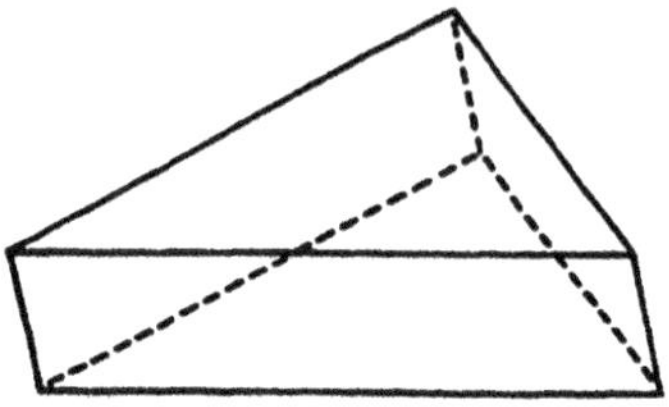

Figure 2.20 The prismatic grain shape typical of a solid with a highly anisotropic surface energy and a hexagonal close-packed crystal structure.

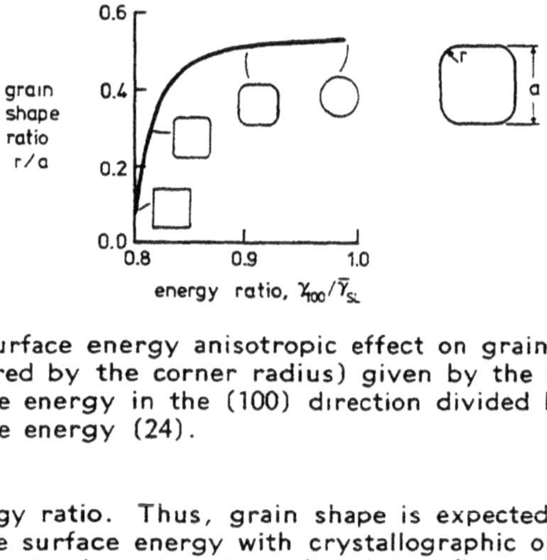

Figure 2.21 The surface energy anisotropic effect on grain shape (as measured by the corner radius) given by the ratio of the surface energy in the (100) direction divided by the mean surface energy (24).

in the surface energy ratio. Thus, grain shape is expected to be sensitive to small changes in the surface energy with crystallographic orientation. Interestingly, the grain shape is not predicted to influence the rate of grain growth (11,25).

H. Mean Grain Separation

The mean grain separation measures the matrix phase thickness between grains. It is important to the mechanical behavior of liquid phase sintered hard materials where the binder phase is ductile. One uncertainty in the past has been whether solid-solid grain contacts constitute a zero mean grain separation. Some reported data ignore the solid-solid grain contacts in measuring the mean grain separation. A general definition based on microscopy depends on the ratio of the volume fraction of liquid V_L and the number of grains per unit line length N_L (9,26),

$$\lambda = V_L/N_L \tag{2.13}$$

where λ is the mean grain separation. The mean random spacing is the mean center-to-center distance. It is the inverse of the number of intercepts per unit line length. The mean grain intercept size L is related to the mean grain separation as follows:

$$L = (1/N_L) - \lambda. \tag{2.14}$$

All of these models assume zero porosity in the microstructure.

In the past it has been stated that the mean grain separation is independent of the grain shape. This is not strictly valid at high volume fractions of solid. As will be shown, the relative amount of solid-solid interfacial area increases rapidly at high volume fractions of solid. Consequently, the solid-solid contacts must be taken into account in considering the mean separation between grains. In general, the mean grain separation varies directly with the mean grain size at a constant volume fraction of solid.

I. Contiguity

Contiguity is a measure of the solid-solid contact in a liquid phase sintered material. In the late stages of liquid phase sintering it is independent of the grain size. The surface area of solid-solid contacts as a fraction of the total interfacial area is termed the contiguity C_{SS},

$$C_{SS} = S_{SS}/(S_{SS} + S_{SL}) \tag{2.15}$$

where the solid-solid surface area per grain is S_{SS} and the solid-matrix surface area per grain is S_{SL}.

The contiguity can be measured metallographically using the number of intercepts per unit length of test line N (9),

$$C_{SS} = 2N_{SS}/(2N_{SS} + N_{SL}) \tag{2.16}$$

where the subscript SS denotes the solid-solid intercepts and SL denotes the solid-matrix intercepts. The factor 2 in Equation (2.16) is necessary since the solid-solid grain boundaries are only counted once by this technique, but are shared by two grains.

In all cases the contiguity increases with increasing volume fraction of solid and dihedral angle. Measurements on several cemented carbides demonstrate this effect. Figure 2.22 shows the measured contiguities of TaC-Co and VC-Co versus the volume fraction of liquid. The dihedral angle for VC-Co is smaller and as a consequence gives a lower contiguity at a given volume fraction of solid. Both systems approach a contiguity of unity as the volume fraction of liquid approaches zero.

For two spheres of equal size, the dihedral angle represents a balance of interfacial energies. Shown in Figure 2.23 are two spheres with a radius of R and a neck between them, as dictated by the dihedral angle. The equilibrium radius of the neck X between the spheres is given as,

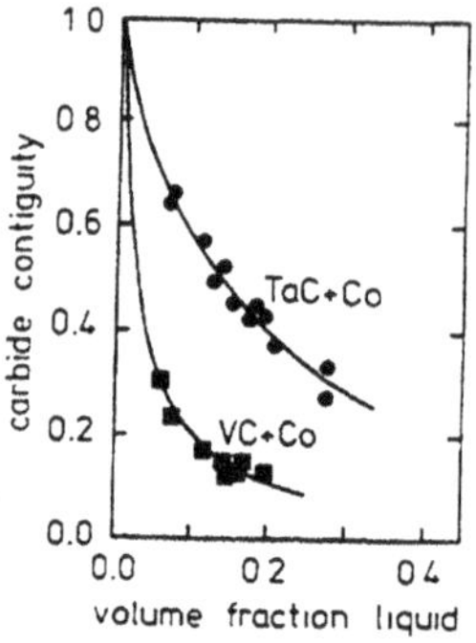

Figure 2.22 Data from Warren and Waldron (7) showing the carbide contiguity variation with volume fraction liquid for two cemented carbides.

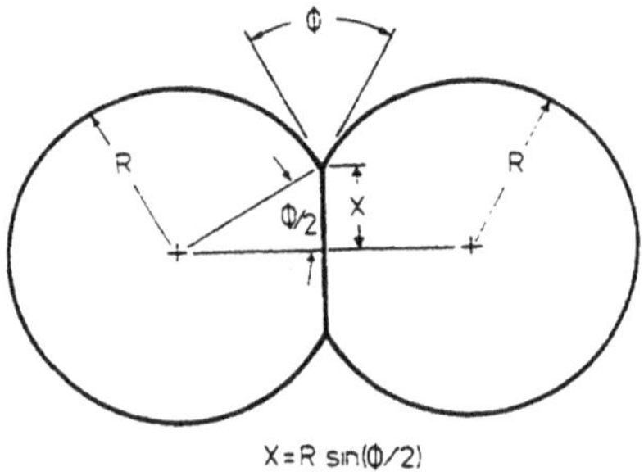

Figure 2.23 A model two grain geometry showing the dihedral angle and neck between the two grains. The neck radius X depends on the grain radius R and the dihedral angle ϕ.

$$X = R \sin(\phi/2). \tag{2.17}$$

Depending on the number of contacts per sphere, the contiguity can be calculated directly from the dihedral angle. The solid-solid contact area is

$$S_{SS} = \pi X^2. \tag{2.18}$$

German (27) has combined the above equations, and expressed the coordination number in terms of the volume fraction of solid, to give the results shown in Figure 2.24. This plot shows the contiguity versus volume fraction of solid for several dihedral angles, exhibiting behavior similar to that noted in Figure 2.22. Since these calculations assume no shape accommodation, the relations are not accurate in the shaded region.

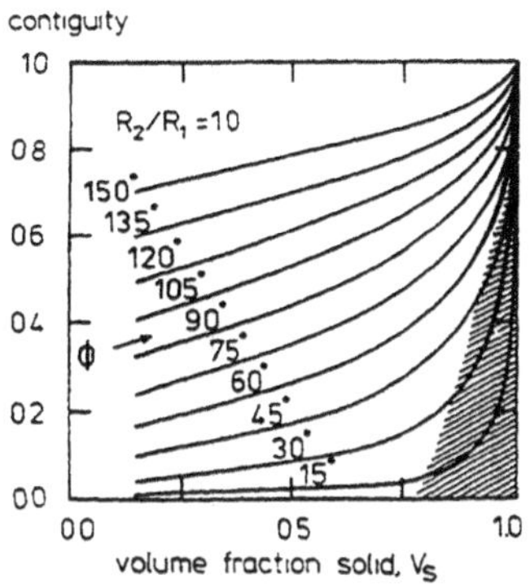

Figure 2.24 The contiguity versus volume fraction of solid for various dihedral angles (27). Within the shaded region the calculations are inaccurate because of grain shape accommodation.

A difference in grain size of contacting grains causes a curved grain boundary between the grains. Even with a range in grain sizes, the dihedral angle holds constant. The radius of curvature of the intergranular boundary depends on the ratio of grain sizes. The boundary is flat for equal sized grains and becomes more curved as the grain sizes differ. Figure 2.25 illustrates the increasing grain boundary curvature with an increasing ratio of grain sizes. In Figure 2.25a the grains are of equal size and the grain boundary is flat. In Figures 2.25b,c,d the radius ratio (smallest grain diameter divided by largest grain diameter) decreases from 0.7 to 0.3. For all of these figures the dihedral angle is held constant at 60°, which corresponds to a surface energy ratio of 1.73. A curved grain boundary provides a driving force for grain coalescence during liquid phase sintering.

Another typical effect of a grain size distribution is to lower the contiguity at a given dihedral angle and volume fraction of solid. Figure 2.26 illustrates the decrease in contiguity associated with such a grain size distribution. In this case the contiguity is shown versus the dihedral angle for a constant volume fraction of solid of 0.8. The behavior for monosized grains is the same as shown in Figure 2.24.

The dihedral angle and volume fraction of solid have a large effect on the contiguity. A change in the energy of either the solid-solid grain boundary or the solid-liquid surface results in a change in the dihedral angle, and a change in the contiguity is therefore expected. Studies on tantalum carbide sintered with a cobalt matrix substantiate these effects (7). Furthermore, because the surface energy changes during the initial portion of liquid phase sintering (see Figure 2.10), the contiguity will change. As noted earlier, chemical reactions associated with melt formation, spreading, penetration of grain boundaries, and dissolution of solid give a varying solid-liquid surface energy. As a consequence, a varying contiguity is expected during the initial portion of liquid phase sintering. Experiments have noted a transient behavior in the contiguity during the first hour of liquid phase sintering (6,7). Figure 2.27 shows an example of this behavior for a W-Ni

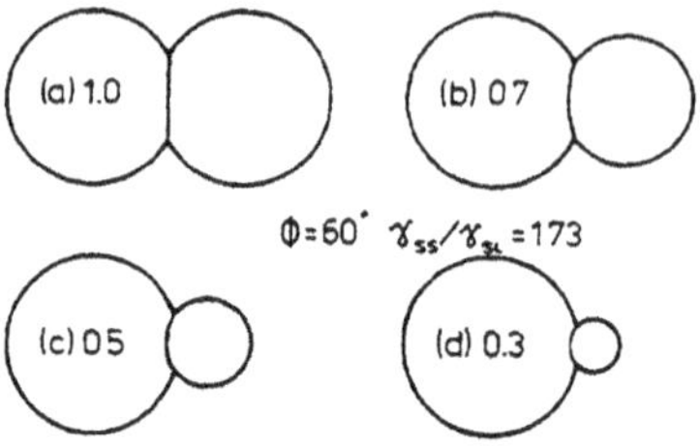

Figure 2.25 The grain boundary curvature at an intergrain neck for equilibrium configurations with a dihedral angle of 60° and grain size ratios of (a) 1.0, (b) 0.7, (c) 0.5 and (d) 0.3.

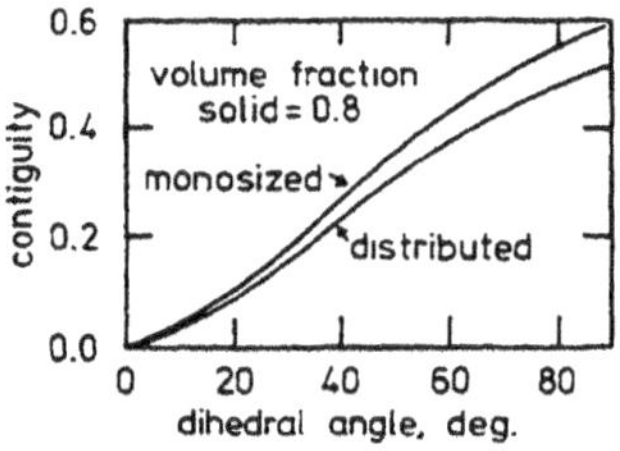

Figure 2.26 The contiguity variation with dihedral angle for a constant volume fraction of solid of 0.8, comparing the effect of a grain size distribution (27).

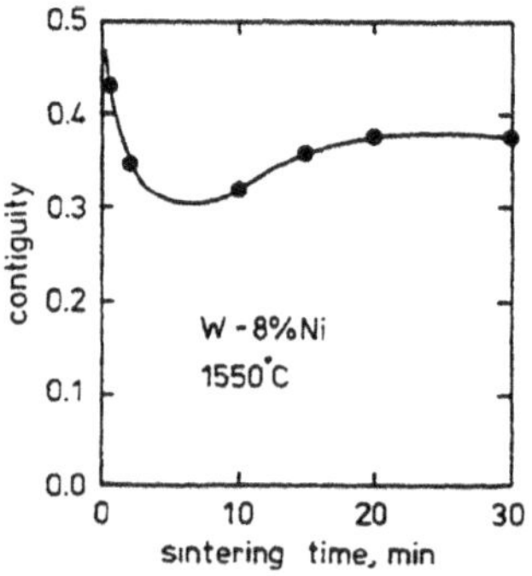

Figure 2.27 The contiguity versus sintering time for a W-Ni compact sintered at 1550°C, showing variations due to surface energy changes during the initial solid-liquid reaction (6).

alloy sintered at 1550°C. The contiguity versus time curve shows a behavior similar to the previously discussed surface energy and dihedral angle time dependencies (Figure 2.10). At long sintering times an equilibrium contiguity is expected.

J. Connectivity

The microstructural connectivity is determined by the number of contacts per solid grain. The solid-solid contacts are stabilized by a nonzero dihedral angle. As a consequence the solid structure forms a rigid skeleton. If the mean number of contacts per grain exceeds 2 during liquid phase sintering, then a rigid structure is expected. The connectivity influences the electrical conductivity, strength, ductility, elastic behavior, dimensional uniformity, and thermal characteristics of material processed by liquid phase sintering. Connectivity and contiguity are positively correlated. At a low volume fraction of solid, contacts form due to gravity induced settling and Brownian motion (28,29). The number of contacts per grain is expected to increase as coalescence and grain shape accommodation increase. Thus, connectivity should increase as the contiguity increases.

A continuous network of solid grains is statistically expected at approximately 36 to 38 volume percent solid. Gurland (30) has measured the electrical conductivity of silver-bakelite mixtures containing various amounts of silver. His results show a drastic decrease in resistivity with a small change in the amount of silver at approximately the critical 36 volume percent solid level. These data are shown in Figure 2.28. Microstructural analysis by Gurland suggests this change in resistivity occurs at approximately 1.4 solid-solid contacts per particle in a metallographic cross section. Prabhu and Vest (31) indicate a change in resistivity at a much lower volume fraction of solid for a ruthenium oxide-glass mixture. No explanation has been offered as yet for this anomalous result.

Contacts per grain increase with the volume fraction of solid. The problem has been addressed in several past studies (32-36). The mean number of contacts per grain depends on the dihedral angle, but typically varies from 14 at a volume fraction of solid equal to 100%, to 3 to 6 contacts at 50%. For low dihedral angles, there will be approximately 8 to 12 contacts per grain at 75% solid. Alternatively, for a high dihedral angle, there will be 4 to 6 contacts per grain at 75% solid. Niemi et al. (37) propose that the number of contacts per grain depends only on the volume fraction of solid, as demonstrated in Figure 2.29. This plot shows the number of contacts per grain (in a two-dimensional cross section) versus the volume fraction of solid for several liquid phase sintering systems. However, a universal curve is not expected since the packing coordination will be limited by a high contiguity. Thus, differences in dihedral angle will modify the basic effect as shown in Figure 2.30 for monosized grains. For comparison, the experimental data for TaC-Co from Grathwohl and Warren (38) are included in Figure 2.30. The behavior predicted for monosized spherical grains is a reasonable model for the experimental system, in spite of a grain size distribution and nonspherical grain shape.

During initial liquid formation, transients in the interfacial energies can have an effect on the connectivity. However, microstructural coarsening does not have a significant effect. Because of the grain size distribution, the

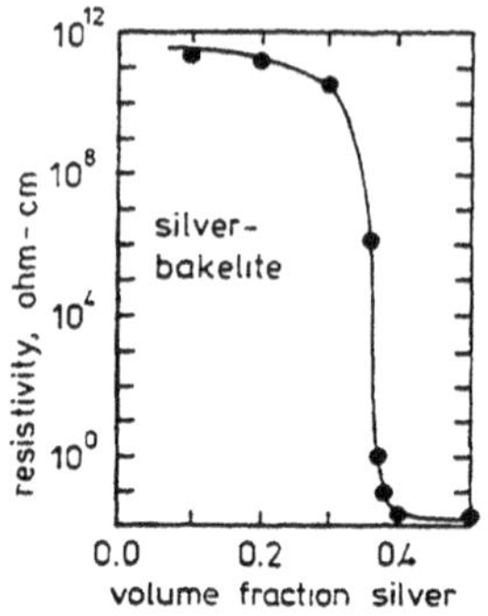

Figure 2.28 Electrical resistivity for a silver-bakelite composite versus the volume fraction of silver. A continuous network of silver occurs a 36 vol.% concentration (30).

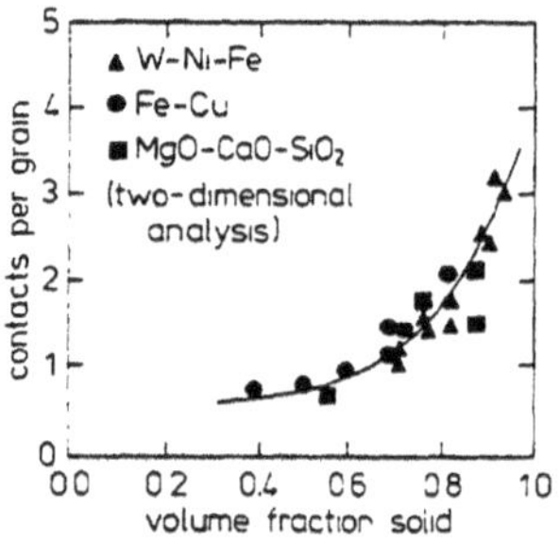

Figure 2.29 The number of contacts per grain versus volume fraction of solid as seen in a two dimensional analysis of various liquid phase sintered materials (37).

microstructure will exhibit a distribution in the number of contacts per grain. The number of contacts per grain will vary by a factor of two within one sample.

K. Neck Size and Shape

Contacts between solid grains during liquid phase sintering grow to a stable size as dictated by the dihedral angle. Earlier it was shown that the equilibrium radius of the grain contact divided by the grain radius (X/R) depended on $\sin(\phi/2)$ (Equation (2.17)). As a consequence the neck size ratio will grow to this size corresponding to the energy minimum (39). The relation given by Equation (2.17) is shown in Figure 2.31 along with some expected grain geometries at dihedral angles of 0, 30, 60, 90, 120, 150, and 180°. Under equilibrium conditions the neck radius X will vary linearly with the grain size. As grain growth occurs, equivalent neck growth also occurs.

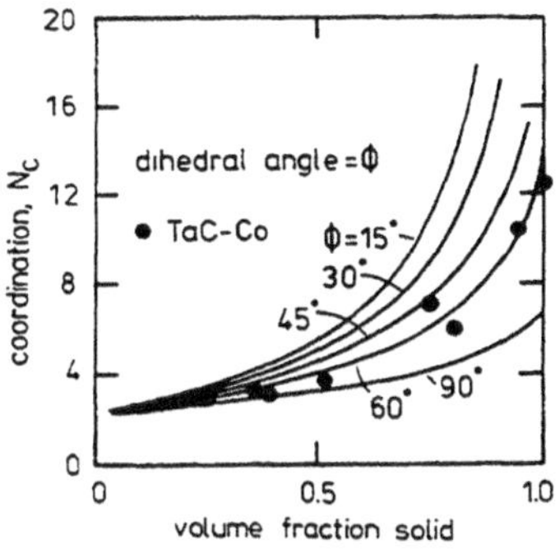

Figure 2.30 The three-dimensional coordination number versus volume fraction of solid for five different dihedral angles (monosized grains). Measured values for TaC-Co (38) are shown for comparison.

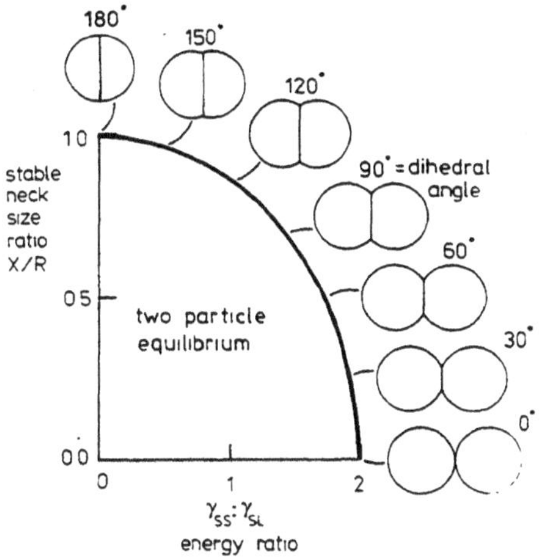

Figure 2.31 The neck size ratio for two contacting grains versus the solid-solid to solid-liquid surface energy ratio.

It is emphasized that this behavior is very different from that encountered in solid state sintering, where the neck size ratio (X/R) continuously enlarge with sintering time.

In general, the contact approaches a circular shape, as can be seen in Figure 2.32. This figure shows a fracture surface for a W-Ni-Fe alloy where the grains are tungsten. The area of the contact can be estimated from

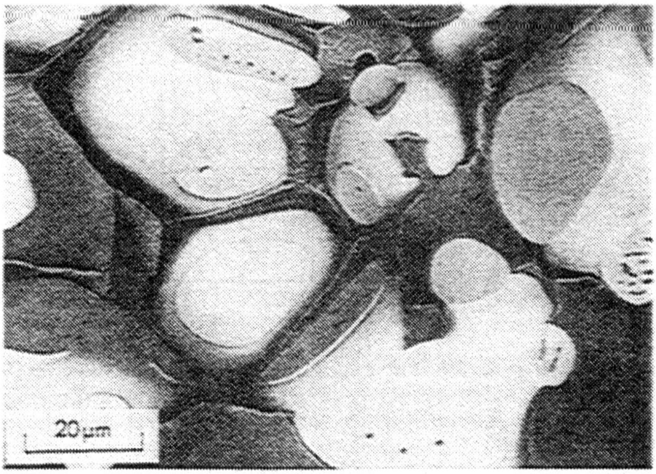

Figure 2.32 A backscatter scanning electron micrograph of a fractured W-Ni-Fe alloy showing the rounded grain shape and circular fractured necks between grains.

Equation (2.17). This treatment assumes isotropic surface energy. If the surface energy is not isotropic, there will be a distribution in the neck size ratio.

As noted earlier, a contact between grains of different sizes produces a curved grain boundary (Figure 2.25). At high temperatures, the grain boundary will migrate, leading to grain coalescence, wherein the large grains absorb the small grains. Coalescence can give some elongated grain shapes and neck sizes, as illustrated in Figure 2.33. Analysis by Makarova et al. (40) suggests that coalescence is a transient event. However, clear experiments have not been performed to identify its overall importance to microstructure development during liquid phase sintering.

L. Summary

This chapter has reviewed the main microstructural concerns and relations associated with liquid phase sintering. Several factors are simultaneously changing during liquid phase sintering. The porosity is usually decreasing, while the grain size is increasing. Furthermore, the dihedral angle and contiguity will vary during the initial portion of liquid phase sintering. These microstructural changes provide a monitor on both the thermodynamics and kinetics during sintering. Later we will employ these microstructural features to describe liquid phase sintered materials, their properties, and their processing.

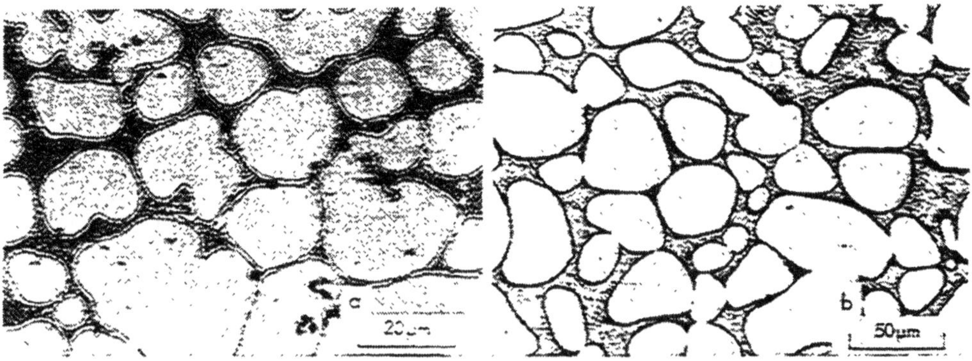

Figure 2.33 Optical micrographs of Fe-Ti (a) and W-Ni-Fe (b) alloys showing coalescence of solid grains (photos courtesy of J. Dunlap and E. Zukas).

An important component of liquid phase sintering is the interfacial energy. The contact angle will depend on the solid-liquid, liquid-vapor, and solid-vapor energies. The dihedral angle will depend on the solid-solid and solid-liquid interfacial energies. These energies depend on solubility, surface contamination, and temperature, and they can change during liquid phase sintering as reactions occur. Furthermore, anisotropic properties are possible. Thus, the microstructural parameters which depend on these energies will be shifting with time during initial liquid phase sintering.

In dealing with liquid phase sintered materials, it must be remembered that initially there are three components (solid, liquid and vapor). The goal of microstructural studies is to describe the amount of each phase, its distribution, and its composition. This requires descriptors of size (grain size, pore size, surface area, mean grain separation), shape (grain shape, pore shape), distribution, and relations between these phases. The relations are developed by the contiguity, connectivity, mean grain separation, and neck size. In general, these various microstructural parameters are interrelated and have a pronounced effect on properties. Beyond the microstructures shown in this chapter, it is possible to have multiple solid phases. In such cases, additional microstructural parameters must be introduced to discuss all of the interactions. The optimization of liquid phase sintered material properties relies on isolation of the parametric influences. This in turn requires careful quantification of the microstructure and its variation with the processing factors.

M. References

1. C. H. Kang and D. N. Yoon, "Coarsening of Cobalt Grains Dispersed in Liquid Copper Matrix," *Metall. Trans. A*, 1981, vol.12A, pp.65-69.
2. S. S. Kang and D. N. Yoon, "Kinetics of Grain Coarsening During

Sintering of Co-Cu and Fe-Cu Alloys with Low Liquid Contents," *Metall. Trans. A*, 1982, vol.13A, pp.1405-1411.
3. J. P. Hirth and J. Lothe, *Theory of Dislocations*, McGraw-Hill, New York, NY, 1968, p.673.
4. O. K. Riegger and L. H. Van Vlack, "Dihedral Angle Measurement," *Trans. TMS-AIME*, 1960, vol.218, pp.933-935.
5. I. A. Aksay, C. E. Hoge, and J. A. Pask, "Phase Distribution in Solid-Liquid-Vapor Systems," *Surfaces and Interfaces of Glass and Ceramics*, V. D. Frechette, W. C. Lacourse and V. L. Burdick (eds.), Plenum Press, New York, NY, 1974, pp.299-321.
6. H. Riegger, J. A. Pask, and H. E. Exner, "Direct Observation of Densification and Grain Growth in a W-Ni Alloy," *Sintering Processes*, G. C. Kuczynski (ed.), Plenum Press, New York, NY, 1980, pp.219-233.
7. R. Warren and M. B. Waldron, "Microstructural Development During the Liquid-Phase Sintering of Cemented Carbides I. Wettability and Grain Contact," *Powder Met.*, 1972, vol.15, pp.166-180.
8. T. H. Courtney, "Densification and Structural Development in Liquid Phase Sintering," *Metall. Trans. A*, 1984, vol.15A, pp.1065-1074.
9. E. E. Underwood, *Quantitative Stereology*, Addison-Wesley, Reading, MA, 1970.
10. R. L. Fullman, "Measurement of Particle Size in Opaque Bodies," *Trans. TMS-AIME*, 1953, vol.197, pp.447-452.
11. O. J. Kwon and D. N. Yoon, "The Liquid Phase Sintering of W-Ni," *Sintering Processes*, G. C. Kuczynski (ed.), Plenum Press, New York, NY, 1980, pp.203-218.
12. O. J. Kwon and D. N. Yoon, "Closure of Isolated Pores in Liquid Phase Sintering of W-Ni," *Inter. J. Powder Met. Powder Tech.*, 1981, vol.17, pp.127-133.
13. H. H. Park, S. J. Cho, and D. N. Yoon, "Pore Filling Process in Liquid Phase Sintering," *Metall. Trans. A*, 1984, vol.15A, pp.1075-1080.
14. R. M. German and K. S. Churn, "Sintering Atmosphere Effects on the Ductility of W-Ni-Fe Heavy Metals," *Metall. Trans. A*, 1984, vol.15A, pp.747-754.
15. D. Kellam and P. S. Nicholson, "Pore Shape Changes During the Initial Stages of Sintering," *J. Amer. Ceramic Soc.*, 1971, vol.54, pp.127-128.
16. R. M. German, "The Direct Observation of Open Porosity Networks," *Metallog.*, 1972, vol.5, pp.462-465.
17. T. K. Kang and D. N. Yoon, "Coarsening of Tungsten Grains in Liquid Nickel-Tungsten Matrix," *Metall. Trans. A*, 1978, vol.9A, pp.433-438.
18. W. Beere, "A Unifying Theory of the Stability of Penetrating Liquid Phases and Sintering Pores," *Acta Met.*, 1975, vol.23, pp.131-138.
19. H. H. Park and D. N. Yoon, "Effect of Dihedral Angle on the Morphology of Grains in a Matrix Phase", *Metall. Trans. A*, 1985, vol.16A, pp.923-928.
20. P. J. Wray, "The Geometry of Two-Phase Aggregates in which the Shape of the Second Phase is Determined by its Dihedral Angle," *Acta Met.*, 1976, vol.24, pp.125-135.
21. R. Raj, "Morphology and Stability of the Glass Phase in Glass-Ceramic Systems," *J. Amer. Ceramic Soc.*, 1981, vol.64, pp.245-248.
22. S. Sarin and H. W. Weart, "Factors Affecting the Morphology of an Array of Solid Particles in a Liquid Matrix," *Trans. TMS-AIME*, 1965, vol.233, pp.1990-1994.
23. L. Murr, *Interfacial Phenomena in Metals and Alloys*, Addison-Wesley, Reading, MA, 1975, pp.3-6.

24. R. Warren, "Microstructural Development During the Liquid-Phase Sintering of Two-Phase Alloys with Special Reference to the NbC/Co System," *J. Mater. Sci.*, 1968, vol.3, pp.471-485.
25. C. K. L. Davies, P. Nash, and R. N. Stevens, "The Effect of Volume Fraction of Precipitate on Ostwald Ripening," *Acta Metall.*, 1980, vol.28, pp.179-189.
26. E. E. Underwood, "Quantitative Stereology for Microstructural Analysis," *Microstructural Analysis*, J. L. McCall and W. M. Mueller (eds.), Plenum Press, New York, NY, 1973, pp.35-66.
27. R. M. German, "The Contiguity of Liquid Phase Sintered Microstructures," *Metall. Trans. A*, 1985, vol.16A, pp.1247-1252.
28. T. H. Courtney, "Microstructural Evolution During Liquid Phase Sintering: Part I. Development of Microstructure," *Metall. Trans. A*, 1977, vol.8A, pp.679-684.
29. T. H. Courtney, "Microstructural Evolution During Liquid Phase Sintering: Part II. Microstructural Coarsening," *Metall. Trans. A*, 1977, vol.8A, pp.685-689.
30. J. Gurland, "An Estimate of Contact and Continuity of Dispersions in Opaque Samples," *Trans. TMS-AIME*, 1966, vol.236, pp.642-646.
31. A. N. Prabhu and R. W. Vest, "Investigation of Microstructure Development in Ruthenium Oxide - Lead Borosilicate Glass Thick Films," *Sintering and Catalysis*, G. C. Kuczynski (ed.), Plenum Press, New York, NY, 1975, pp.399-408.
32. F. B. Swinkels and M. F. Ashby, "A Second Report on Sintering Diagrams," *Acta Met.*, 1981, vol.29, pp.259-281.
33. J. P. Jernot, J. L. Chermant, and M. Coster, "The Mean Free Path in the Porous Phase of Sintered Materials," *Powder Tech.*, 1981, vol.30, pp.31-35.
34. R. M. German and Z. A. Munir, "Morphology Relations During Surface-Transport Controlled Sintering," *Metall. Trans. B*, 1975, vol.6B, pp.289-294.
35. S. Prochazka and R. L. Coble, "Surface Diffusion in the Initial Sintering of Alumina. Part I - Model Considerations," *Sci. Sintering*, 1970, vol.2, pp.1-18.
36. H. F. Fischmeister and E. Arzt, "Densification of Powders by Particle Deformation," *Powder Met.*, 1983, vol.26, pp.82-88.
37. A. N. Niemi, L. E. Baxa, J. K. Lee, and T. H. Courtney, "Coalescence Phenomena in Liquid Phase Sintering - Conditions and Effects on Microstructure," *Modern Developments in Powder Metallurgy*, vol.12, H. H. Hausner, H. W. Antes, and G. D. Smith (eds.), Metal Powder Industries Federation, Princeton, NJ, 1981, pp.483-495.
38. G. Grathwohl and R. Warren, "The Effect of Cobalt Content on the Microstructure of Liquid-Phase Sintered TaC-Co Alloys," *Mater. Sci. Eng.*, 1974, vol.14, pp.55-65.
39. I. M. Stephenson and J. White, "Factors Controlling Microstructure and Grain Growth in Two-Phase and in Three-Phase Systems," *Trans. Brit. Ceramic Soc.*, 1967, vol.66, pp.443-483.
40. R. V. Makarova, O. K. Teodorovich, and I. N. Frantsevich, "The Coalescence Phenomenon in Liquid Phase Sintering in the Systems Tungsten-Nickel-Iron and Tungsten-Nickel-Copper," *Soviet Powder Met. Metal Ceram.*, 1965, vol.4, pp.554-559.

CHAPTER THREE

Thermodynamic and Kinetic Factors

A. Surface Energy

Surface energies are the major factors determining behavior during liquid phase sintering. The minimum criteria for successful liquid phase sintering are i) a low temperature liquid, ii) solubility of the solid in the liquid, and iii) liquid wetting of the solid grains (1). These conditions result in a reduction in surface energy with liquid spreading. At high volume fractions of solid, the elimination of porosity and its associated surface energy requires shape accommodation on the part of the solid grains, which is dependent on solubility of the solid in the liquid. Furthermore the rate of microstructural coarsening during liquid phase sintering, as seen by the grain growth rate, increases with the solid-liquid surface energy. These factors lead to the conclusion that surface energy is the major driving force for densification.

The surface energy is the work needed to expand a surface normal to itself. Inherently, surface energy is traced to interatomic forces. Consequently, material parameters like the heat of vaporization, hardness, elastic modulus, and melting temperature provide a rough gauge of the surface energy (2).

To better understand surface energies, contrast the interior and exterior regions of a solid. The interior is uniform with homogeneous bonding. Thus, the time averaged net force on an atom is zero. In contrast, the exterior surface represents an unbalanced force situation. The bonding is not homogeneous and the time averaged net force is directed towards the interior. As a consequence, the exterior regions move toward a minimized surface area with a lower surface energy. For a fluid, the liquid-vapor surface energy is uniform in all directions. Thus, for simple liquids, surface energy is characterized by a single value which decreases with increasing temperature. The liquid-vapor surface energy γ_{LV} dependence on temperature T is given as follows (3,4):

$$\gamma_{LV} = \gamma_o \, (1 - T/T_c)^n \tag{3.1}$$

where the exponent n is approximately 1.2, and T_c and γ_o are material parameters.

The effect of surface energy in liquid phase sintering is to drive the microstructure towards a minimum energy configuration. The first step in this process is particle rearrangement when the liquid forms and wets the solid structure. The second step occurs when solution-reprecipitation becomes active, leading to pore elimination by particle shape adjustment. Finally, the third step is seen as microstructural coarsening, where the mean grain size continuously increases, giving a decreasing surface area per unit volume. Inherently, all three of these steps are linked to the surface energy. The Young (Laplace) equation explains much of the microstructural variation observed during liquid phase sintering. The pressure excess ΔP across a curved liquid surface depends on the two radii of curvature (R_i and R_j) and the surface energy as given below,

$$\Delta P = \gamma \, (1/R_i + 1/R_j). \tag{3.2}$$

By convention, if the radius of curvature is located inside the liquid, then the sign is positive. The minimum pressure across a surface occurs when the two radii are equal and opposite in sign, which corresponds to a saddle surface. For a spherical droplet the radii are equal and the same, thus Equation (3.2) becomes

$$\Delta P = 2 \, \gamma / \, R \tag{3.3}$$

with R equal to the droplet radius. According to Equation (3.3), the pressure inside a spherical droplet is greater than the external pressure by a value of ΔP. This spherical shape corresponds to the definition of a minimum surface area liquid (with a constant and isotropic surface energy).

One result from Equation (3.2) is that two coalescing grains of differing size will have an effective pressure difference across the grain boundary. The larger grain will have the lower pressure and be more stable. Therefore, the smaller grain shrinks while the larger grain grows. The process of coalescence is shown diagrammatically in Figure 3.1. The grain boundary between the two grains moves into the smaller grain because of the differing internal pressures. For both concave and convex surfaces the tendency is towards a flat surface. The pressure difference given by Equation (3.2) is the basis for capillary flow of a liquid as discussed later in this chapter.

Another consequence of Equation (3.2) is that a wetting liquid exerts a force on the solid grains. The pressure inside the liquid shown in Figure 3.2 is less than the external pressure. The pressure difference ΔP depends on the liquid-vapor surface energy γ_{LV} as follows:

$$\Delta P = (2 \, \gamma_{LV} \, \cos\Theta)/d \tag{3.4}$$

where Θ is the contact angle and d is the separation between grains. Note

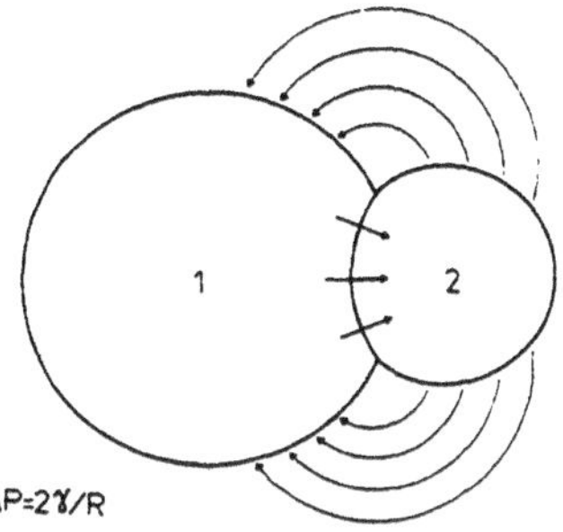

Figure 3.1 Grain coalescence during liquid phase sintering occurs because the larger grain is more stable, leading to growth by grain boundary migration and solution-reprecipitation.

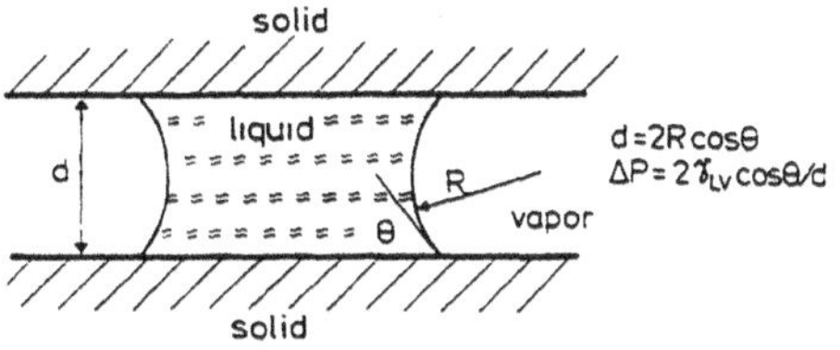

Figure 3.2 An adhesion pressure is exerted by a wetting liquid because of the lower pressure inside the liquid, giving an attractive force between the solid surfaces.

that the pressure difference causes attraction of the grains if the contact angle is less than 90°, while it is repulsive if the contact angle exceeds 90°. Densification in the rearrangement stage of liquid phase sintering is enhanced by a low contact angle and a large liquid-vapor surface energy.

B. Wetting

A natural concern in liquid phase sintering is good wetting. As developed in Chapter 2, good wetting depends on a low solid-liquid surface energy in comparison to the solid-vapor and liquid-vapor surface energies. Typically wetting is best when there is a chemical reaction at the solid-liquid interface (5). Because of such a surface reaction, the reactive metals wet most metal oxides, while the noble metals do not. Thus, the wettability of liquid metals on oxides increases with the oxygen affinity of the liquid metal. Likewise, wettability will decrease as the oxide stability increases. This suggests an inverse relationship between the contact angle and the free energy of oxide formation (6).

There is considerable interest in the metal bonded diamond composites. Best wetting of the diamond powder is obtained when the liquid melt reacts with the diamond. Usually metals with unfilled d or f electron shells are most suitable; titanium, zirconium, chromium, manganese, vanadium, and niobium. Alternatively, poor wetting is obtained from the metals with filled d or f electron shells; copper, silver, and gold.

Similarly, the borides exhibit a wetting dependence on the chemical interaction between the solid and liquid. The more stable the solid boride (the more negative the free energy of formation), then the higher the contact angle between the boride and any given metal. For example, titanium boride is more stable than molybdenum boride. In the presence of a molten metal like nickel, titanium boride has a larger contact angle than the molybdenum boride and nickel combination.

For the metal-metal systems, clean surfaces are important. The lack of chemical interaction as evident in the phase diagram is a good indicator of poor wetting. Fluxes are available for improving wetting. The primary function of a flux is to decompose surface oxides which can inhibit wetting in an otherwise favorable system.

C. Spreading

Spreading is the kinetic process associated with wetting. The spreading of a liquid over a solid is important to the early portion of liquid phase sintering. The spreading liquid can separate intergranular bonds formed during heating (1,7). Thus, liquid spreading leads to a more homogeneous microstructure and can affect the degree of rearrangement.

Study of the spreading of liquids over solid surfaces is difficult. There are several factors which influence the behavior, such as crystallographic orientation, segregation, contamination, and mechanical agitation. In addition, exposure of the solid to the liquid changes the solid surface and contact angle.

A general theory of spreading is lacking. It appears spreading phenomena are specific to individual systems and are highly dependent on the atomic structure and not the bulk properties. Chemical affinity is important to spreading (5). Furthermore, solubility between the two phases aids spreading. It is clear that spreading depends on a reduction in the free energy. For the situation shown in Figure 3.3, spreading of the liquid phase over the solid gives an increase in the liquid-vapor and solid-liquid surface areas and a decrease in the solid-vapor surface area. For this process to occur, the surface energy must be lower after spreading. This requires that,

$$\gamma_{LV} - \gamma_{SV} + \gamma_{SL} < 0 \qquad (3.5)$$

where γ is the surface energy and the subscripts denote the interface (SL = solid-liquid, LV = liquid-vapor and SV = solid-vapor).

After initial spreading, the uniform layer of liquid between two solid grains often decomposes into lens shaped regions. This occurs because of the increasing surface energy which accompanies completion of a reaction across the solid-liquid interface (see Chapter 2). Figure 3.4 shows a schematic diagram of this process and also shows a micrograph of a sintered material

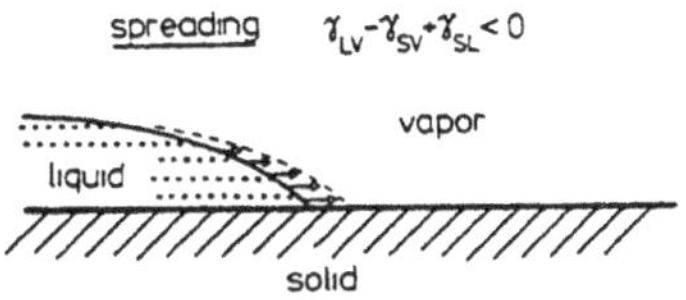

Figure 3.3 The spreading of a liquid over a solid surface. For spreading to be favorable, the net interfacial energy must be reduced.

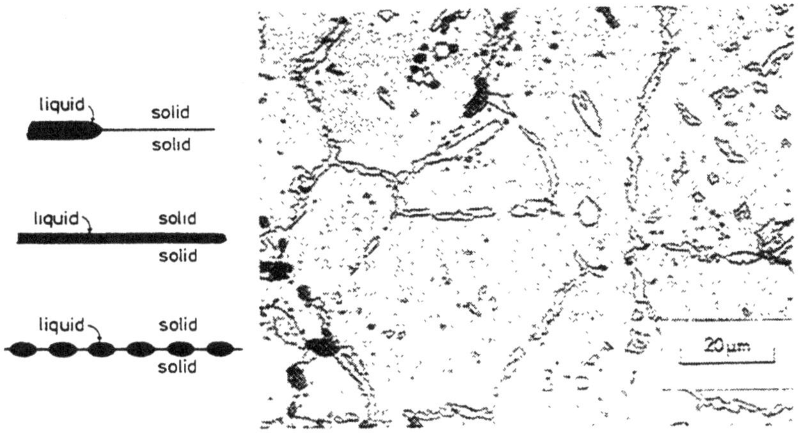

(a) (b)

Figure 3.4 The formation of a necklace structure along grain boundaries due to the penetration of a wetting liquid (a) and an example microstructure corresponding to Fe-7% Ti (b) (photo courtesy of W. H. Baek).

with such a structure along the grain boundaries. Calculations for film breakup into this lens structure show it is stable for all dihedral angles over 0° (8). The lens size depends on both the dihedral angle and volume fraction of liquid. The number of second phase clusters per unit grain boundary area increases as the dihedral angle increases. At small volume fractions of liquid, the structure is composed of small, closely spaced clusters as illustrated in Figure 3.5. This figure shows the effect of variations in dihedral angle and volume fraction of liquid on the resulting microstructure. An important finding from the calculations is that the fractional grain boundary area covered by the second phase after liquid penetration is independent of the volume fraction of liquid and has a negligible dependence on dihedral angle. The estimated fractional coverage at equilibrium after liquid penetration and decomposition is 0.65. Such a model only applies to small volume fractions of liquid such as is evident in Figure 3.4.

In liquid phase sintering several barriers can exist to inhibit spreading. The spreading rate is governed by the mechanism of spreading,

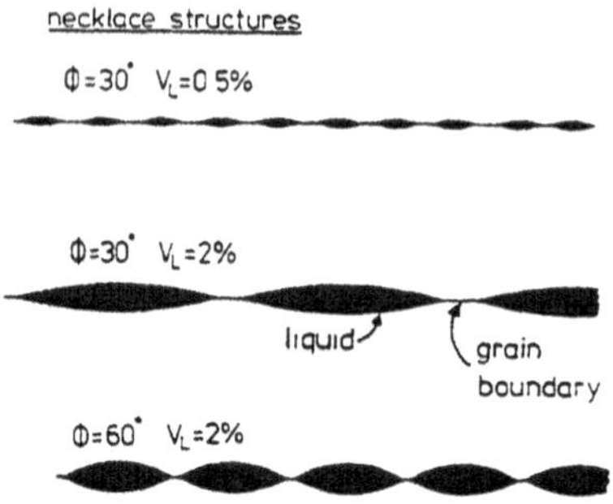

Figure 3.5 Calculated equilibrium lens structures along an equivalent segment of grain boundary for dihedral angles of 30 and 60° and volume fraction of liquid of 0.5 and 2.0%.

and not the free energy decrease associated with wetting. Thus, small concentrations of impurities can greatly alter the spreading behavior. Impurities alter spreading by affecting the kinetics and often do not change the surface energy. In non-spreading systems, small quantities of active agents can induce spreading. In such cases, selective adsorption at the solid-liquid interface reduces the surface energy. Accompanying the surface area reduction is an expansion of the solid-liquid contact area and a contraction in the solid-vapor contact area.

The mechanism of spreading is associated with the chemical affinity and bond formation at the liquid-solid interface. Cohesion of the liquid to the solid causes the liquid to closely follow behind the spreading interface. Individual atomic jumps at the liquid-solid interface usually precede the actual extension of the liquid. Thus, the essential kinetic step is related to diffusion of the liquid over the solid surface by surface diffusion or evaporation-condensation. Consequently, spreading kinetics depend on the underlying surface mobility of the solid. For the early stage of liquid spreading the rate is dependent on the arrival of spreading complexes to the interface. Since spreading is dependent on short range events it is observed that the rate dA/dt varies as follows (5):

$$dA/dt \sim A^n \tag{3.6}$$

where A is the covered area and n is typically between 0.5 and 1.0. The spreading of a liquid over a solid begins as soon as the liquid is formed. During spreading, a reaction occurs at the interface until the liquid approaches saturation with the solid component. The spreading process continues to a point where the contact angle is reached. If the liquid is at an equilibrium composition, then spreading stops. The lower the equilibrium contact and dihedral angles, the greater the extent of liquid spreading throughout the powder compact.

D. Segregation Effects

Impurities segregate to a surface if they lower the free energy. With multiple component liquids, as is typical for liquid phase sintering, the surface layer will become enriched in the lower surface energy component. Such segregation (or selective adsorption) is concentration, temperature, and diffusion dependent. Initially, segregation is rapid, followed by a slow aging process which continually changes the surface energy. For an alloy, the underlying chemistry will have a large influence on the surface chemistry. In dilute solutions, the more a solute tends to lower the interfacial energy, the greater the degree of segregation (5). Figure 3.6 shows the surface energy reduction possible in dilute solutions; in this case, for dilute concentrations of oxygen and sulfur in iron. The surface energy decreases as the concentration of additive increases due to selective segregation to the surface. This segregation effect is usually limited to the first few atomic layers of an interface. In ionic crystals, separate surface phases may form due to extensive surface segregation (9).

There is a general correspondence between surface and grain boundary segregation under equilibrium conditions. In both cases, the relative magnitude of the segregation effect varies with the inverse solid solubility. As an example, the grain boundary enrichment depends on the bulk concentration (10),

$$C_B/(C_S - C_B) = [C_V/(1 - C_V)] \exp(Q/kT) \tag{3.7}$$

where Q is the energy associated with grain boundary segregation, and C is the concentration with the subscript B representing the grain boundary concentration of solute, S the saturation concentration of solute on the boundary, and V the bulk concentration of solute. In Equation (3.7), kT is the product of the Boltzmann constant and the absolute temperature. Generally, the equilibrium saturation value is below 1.5 atomic monolayers. Equation (3.7) shows that the equilibrium segregation increases with decreasing temperature. For most systems, the solubility at saturation for a bulk phase is raised by a higher temperature, with an Arrhenius temperature dependence (4).

The phase diagram provides an initial indicator of segregation and surface energy changes due to alloying and impurities (11). Figure 3.7 shows the basic distinction between segregating and nonsegregating systems. A dilute alloy at the solidus temperature will show enrichment of the solute in the liquid. The greater the separation of the solidus and liquidus curves, the greater the solute segregation to interfaces. Figure 3.7a shows the liquidus and solidus curves associated with no segregation. In Figure 3.7b the downward sloping liquidus and solidus indicate a tendency for solute segregation and lower surface energies. Finally, Figure 3.7c corresponds to a situation where extensive solute segregation is expected. The favorable factors for solute segregation are decreasing liquidus and solidus, wide liquidus and solidus separation, small solute solid solubility, and small atom size.

Solute segregation to grain boundaries and solid-liquid interfaces are similar in many systems (11,12). This is illustrated in Figure 3.8 for tin

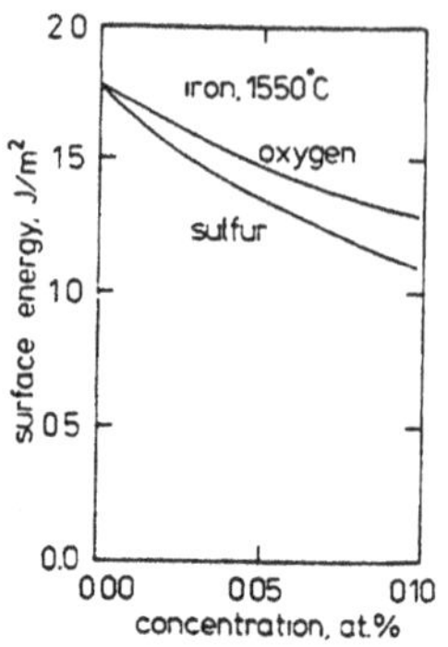

Figure 3.6 The liquid-vapor surface energy for dilute iron alloys with oxygen and sulfur, showing the large effects of small concentrations of additives (10).

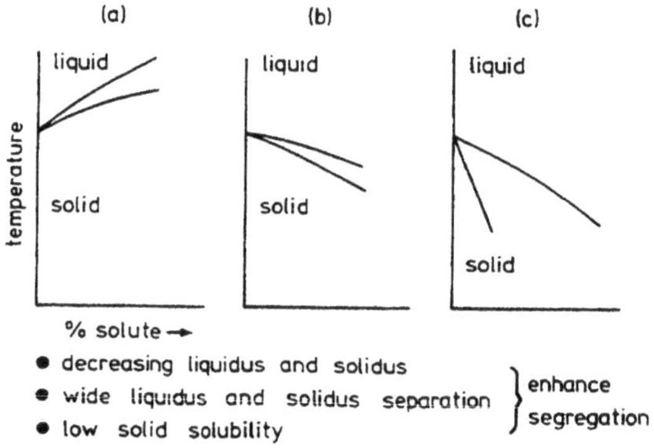

Figure 3.7 Interfacial segregation as indicated by phase diagram features (11). In equilibrium, solute segregation to interfaces will be least for (a) and greatest for (c).

segregation to both grain boundaries and free surfaces of iron at 550°C. One atomic monolayer indicates the interface is essentially pure solute. In cases of preferential segregation there are corresponding decreases in interfacial energies (13), such as illustrated in Figure 3.6. Likewise, as the equilibrium bulk solubility decreases, the amount of segregation to grain boundaries increases. This effect is demonstrated in Figure 3.9 using data for copper, nickel, and iron alloys (14). The enrichment factor is the ratio of the grain boundary concentration to the bulk concentration. The lower the solid solubility the greater the relative degree of segregation. This results in a rapid decrease in the grain boundary energy. For example, in the Cu-Bi system approximately 0.25 atomic percent of bismuth reduces the grain boundary

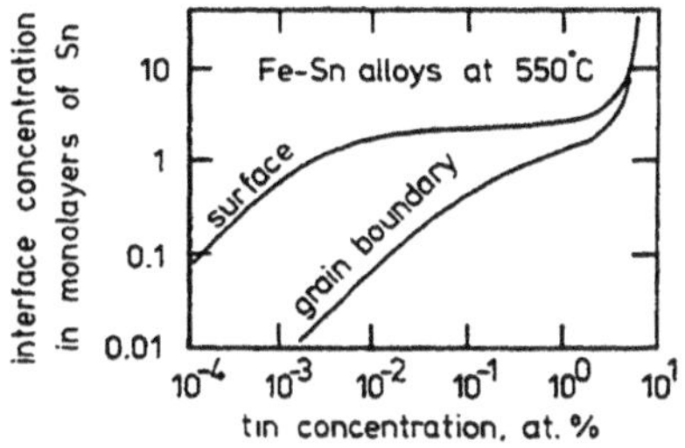

Figure 3.8 Equilibrium segregation of tin to grain boundaries and surfaces of iron at 550°C versus the alloy tin content (12).

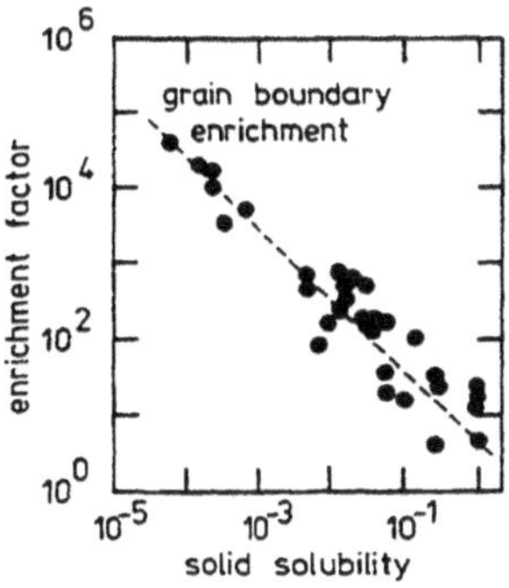

Figure 3.9 Enhanced grain boundary segregation of solute (grain boundary concentration divided by bulk concentration) shown as a function of the atomic fraction solid solubility for Cu, Ni, and Fe alloys (14).

energy of copper from 0.60 to 0.25 N/m (12). Such segregation can have a large effect on liquid phase penetration of grain boundaries. For example, in liquid phase sintered iron-copper-carbon alloys, carbon inhibits penetration of iron grain boundaries by molten copper (15,16). Thus, there is a major difference in the sintering behavior of Fe-Cu versus Fe-Cu-C compacts. The Fe-Cu compositions swell during the initial portion of sintering due to liquid penetration of grain boundaries. Alternatively, the Fe-Cu-C compositions show less swelling because of preferential carbon diffusion to the grain boundaries before the copper melts. Any change which alters an interfacial energy can also be expected to effect the other system parameters such as dihedral angle and contiguity. These system parameters are variable during the reaction portion of liquid phase sintering (17-20).

Impurity segregation has a large effect on liquid phase sintering. In the metallic systems, liquid spreading and grain boundary penetration are highly dependent on the segregated species. In ceramic systems there is evidence for liquid phase sintering due to impurity concentration at interfaces (21). Indeed, this phenomenon appears to be widespread in technical ceramics with as little as a few hundred parts per million of impurities. Transmission electron microscopy studies after sintering show glassy phases composed of impurities concentrated at grain boundaries. The glassy phase composition corresponds to the expected liquid phase at the sintering temperature.

Another aspect of segregation is with respect to the dihedral angle. Confusion exists as to the true dihedral angle in systems like WC-Co. The dihedral angle measured from optical micrographs is much larger than zero. However, transmission electron microscopy shows cobalt layers at the carbide grain boundaries. The confusion from such examinations is how to reconcile the nonzero dihedral angle with the cobalt layers along the solid-solid contacts. From the above discussion on segregation it is evident that such behavior results from segregation during cooling. During sintering the dihedral angle is nonzero, while during cooling cobalt segregates to the interface due to a decreasing solubility in the carbide. This segregation leads to the impression that liquid was present at the interface during sintering.

With initial liquid formation, there is often a rapid decrease in the solid-liquid surface energy. Thus, there is rapid liquid penetration of grain boundaries and a drastic microstructural change. This is the precursor condition to the formation of the necklace microstructure described earlier. During this portion of liquid phase sintering the penetration of the liquid can be as fast as 2 μm/s. Rapid spreading of the liquid is observed in most of the successful liquid phase sintered materials. Under non-equilibrium conditions, the initial stage reaction at the solid-liquid interface leads to mass transfer and a substantially reduced surface energy. The decrease in surface energy depends on the free energy per unit area of the chemical reaction. Reactions involving soluble components aid wetting and cause spreading on the solid. During liquid spreading several microstructural changes take place and there can even be changes in the anisotropic nature of surface energies (22). Experiments by Whalen and Humenik (23) show that the contact angle varies with the liquid composition in Fe-C liquids on graphite. Figure 3.10 shows their results for the contact angle versus the amount of carbon in the liquid. The solid-liquid surface energy is low as long as carbon is diffusing across the reacting interface. When the equilibrium value of carbon in iron is obtained, poor wetting is observed. Similarly, in cemented carbides like TiC-Ni, the addition of molybdenum initially aids wetting. The attainment of equilibrium is delayed because of slower interfacial transport with molybdenum present in the liquid. Liquid phase sintering improves with better wetting, more liquid phase, and increased solid solubility in the liquid. All of these factors are favored by higher sintering temperatures in most systems. Segregation on cooling depends on the cooling rate, solubility, and diffusivity; thus, the final atomic scale microstructure will be cooling rate sensitive.

E. Capillarity

Another aspect of interfacial energies associated with liquid phase sintering is capillarity. A thin capillary tube and a wetting liquid lead to the phenomenon known as capillary rise. The height the liquid penetrates up the tube against gravity depends on the contact angle, tube diameter, and

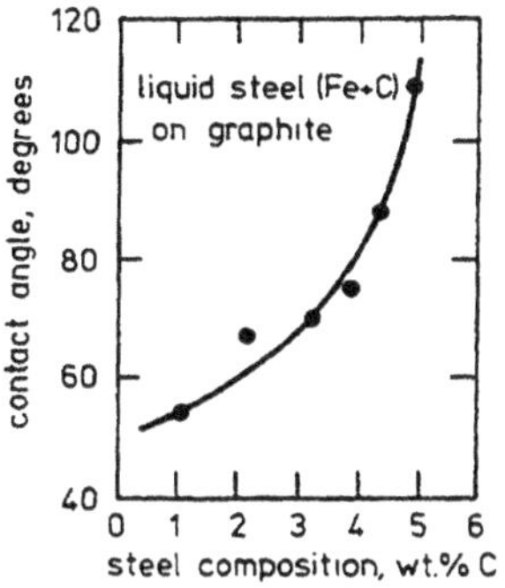

Figure 3.10 The contact angle between molten steel and graphite for various steel carbon contents during diffusion across the interface (23).

interfacial energy. The capillary attraction is increased by a smaller capillary tube diameter. In liquid phase sintering this means that a wetting liquid will preferentially flow into the region of smallest pore diameter. The typical green compact microstructure is inhomogeneous; differences in powder packing and pore size lead to gradients in liquid penetration. If the liquid forming additive has a relatively large particle size, then on melting it will flow into the neighboring pores and create a large pore at the prior additive particle site (24). These pores are easier to remove in subsequent sintering if the additive has a small particle size. Thus, smaller additive particle sizes are beneficial because of smaller pores, better homogeneity, and better capillarity.

A beneficial aspect of capillarity is the strong surface force exerted on the solid when the liquid forms. The attractive force between two particles with a wetting liquid causes rearrangement, densification and contact flattening (25). Using the two sphere model shown in Figure 3.11, the pressure difference between the liquid meniscus and the vapor is given by Equation (3.2). The radii of curvature of the liquid depend on the amount of liquid, contact angle, particle separation, and particle sizes. The force between the two spheres in Figure 3.11 is given by summing the pressure and surface energy contributions (26-30). The force F on one of the particles depends on the meniscus size X as follows:

$$F = \pi X^2 \Delta P + 2 \pi X \gamma_{LV} \cos\Psi \qquad (3.8)$$

where ΔP is the pressure given by Equation (3.2), and Ψ is the angle shown in Figure 3.11. At equilibrium the energy of the configuration must be at a minimum. To determine the equilibrium condition, it is necessary to know the liquid profile. Generally, this profile is determined by numerical solution to a governing differential equation (30). It should be noted that the true solution is a nodoid and is not a circular arc as has been assumed in some cases (31,32).

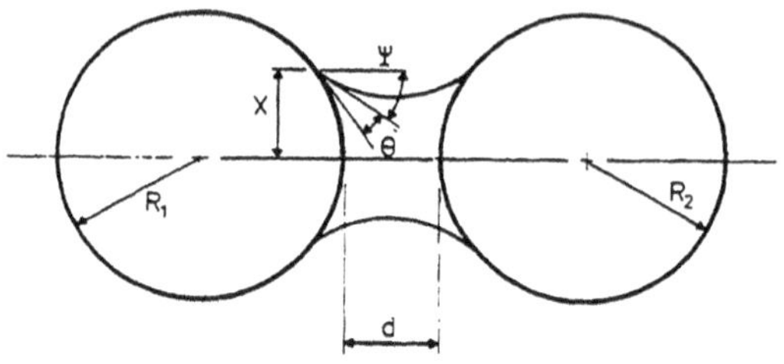

Figure 3.11 Two spherical particles with a liquid bridge and the geometric factors involved in calculating the wetting force.

For good wetting, the force is attractive, leading to zero separation between the particles. Alternatively, for a poor wetting liquid, the liquid forces a separation of the particles. These two conditions are contrasted in Figure 3.12. In Figure 3.12a the low contact angle gives a tensile (attractive) force between the two particles. Alternatively, in Figure 3.12b the high contact angle gives repulsion between the particles. In the extreme of 0° contact angle, all separations between the particles are resisted for all particle sizes and amounts of liquid. A combination of Equations (3.2) and (3.8) shows the attractive force due to a wetting liquid varies with the surface energy. Figure 3.13 shows the results from experiments by Eremenko et al. (6) involving 4 mm diameter steel balls and three different wetting liquids. Note the attractive force decreases as the liquid surface energy decreases and as the quantity of liquid increases.

If contacting particles with a wetting liquid are separated, then the liquid bridge eventually breaks apart as illustrated in Figure 3.14. In this case the restoring force is lost (33). Figure 3.15 plots the attractive force between contacting spherical grains versus the separation distance for a contact angle of 0°. The plot includes two volume fractions of liquid (1 and 10% of the smaller particle volume) and two particle size ratios (1:1 and 1:5). For this plot the force has been normalized by dividing by the particle radius and the liquid-vapor surface energy. (A combination of Equations (3.2) and (3.8) for the liquid-solid contact point shows the origin of this normalization.) Furthermore, the separation is also normalized to the particle size of the smaller particle. From Figure 3.15 it can be seen that the contact force for a wetting liquid decreases with particle separation. Thus, particles with a connecting meniscus of wetting liquid will be drawn together with an increasing force. Equilibrium corresponds to zero separation. Furthermore, smaller particle sizes are beneficial since the magnitude of the attractive force is inversely dependent on the particle size. For many practical systems, the particle diameter must be below 10 μm for the capillary forces to give significant densification by rearrangement during liquid phase sintering.

The capillary force for poor wetting (θ = 85°) is shown in Figure 3.16 for the same volume fractions of liquid and particle size ratios as shown in Figure 3.15. Again the dimensions are normalized. In this case there is an equilibrium particle separation for the smaller quantities of liquid. Particles

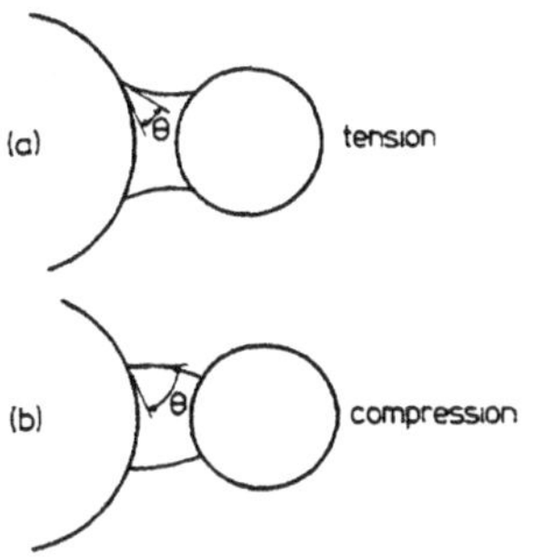

Figure 3.12 The effect of the two extremes of contact angle; (a) good wetting, leading to an attractive force, and (b) poor wetting, leading to a repulsive force.

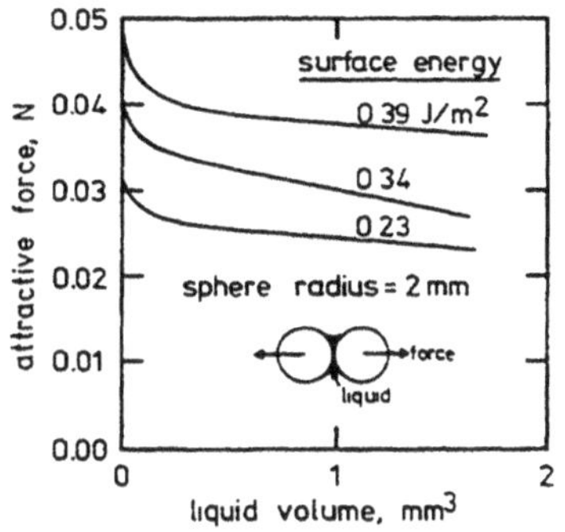

Figure 3.13 The attractive force between two steel balls shown as a function of the liquid volume for three different wetting liquids (6).

with a separation larger than this equilibrium value will be drawn together, while particles with lesser separations will be repulsed. The magnitude of the equilibrium separation increases with the contact angle, volume fraction liquid, and decreases with differences in particle sizes. In contrast, the wetting liquid shows an increasing force with no separation as the liquid volume decreases and as the particle sizes differ. Obviously, there is a transition condition where the force is zero for various separation distances, contact angles, and volume fractions of liquid. Figure 3.17 shows the relation between these parameters. This is a plot of the contact angle and normalized volume fraction liquid corresponding to zero interparticle force for monosized particles and a 5 to 1 particle size ratio. For low contact angles (below approximately 60°) there will always be attraction between particles. Comparison with experimentally measured attraction forces between wetted spheres shows these models are accurate (6,30). Generally, it is safe to

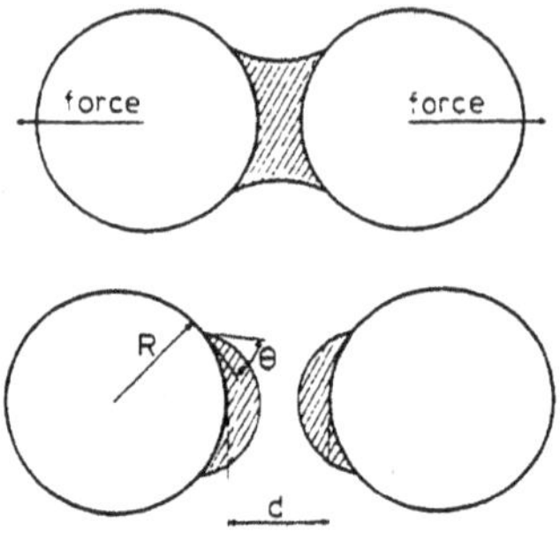

Figure 3.14 Rupture of the wetting liquid bridge between two solid spheres due to an applied external force.

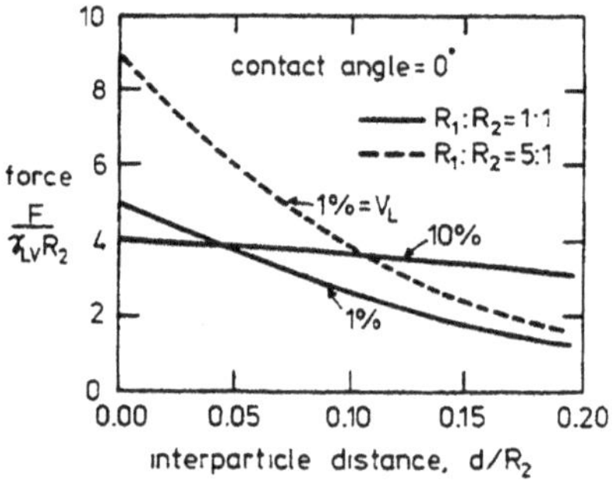

Figure 3.15 The attractive force between two spheres with a wetting liquid bridge versus the separation distance for two ratios of particle size and liquid contents of 1% and 10% of the smaller sphere volume (30).

ignore gravity effects in the analysis of capillary forces for liquid phase sintering systems. For example, for a particle size of 1 μm and a typical liquid-vapor surface energy and 5 vol.% liquid, the capillary force is 10,000 times the gravitational force.

It is significant to note the universal nature of the normalized force curves (34). The force pulling the particles together goes down with an increasing volume fraction of solid. However, the force has been found to drop suddenly to zero as the volume fraction of liquid approaches zero (6). For typical liquid phase sintering, the pores are submicron in size; thus, the capillary forces causing rearrangement are quite strong. For a wetting liquid, the contact angle is near 0° and the liquid will occupy the smallest pores. For a nonwetting liquid, the contact angle is large and the capillary force leads to the liquid occupying the largest pores. These differences can be seen in sintered microstructures such as those shown in Figure 2.2.

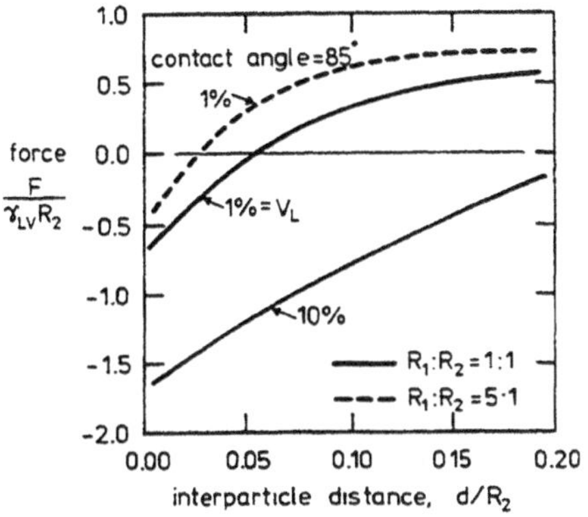

Figure 3.16 The normalized force between two spheres with a liquid bridge at a contact angle of 85°. At small separations the force is repulsive (30).

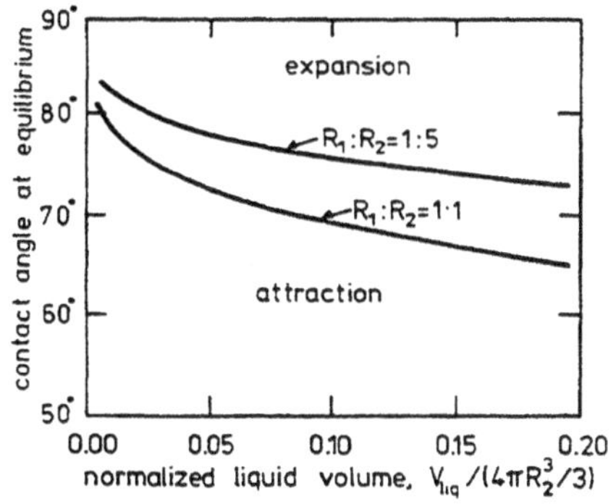

Figure 3.17 The relation between contact angle and volume fraction liquid corresponding to attraction or separation for two size ratios. At low contact angles all quantities of liquid give zero separation (30).

F. Viscous Flow

The capillary force on solid particles due to a wetting liquid causes rearrangement. A high level of rearrangement depends on a low wetting angle, small particle sizes, and minimal solid-solid contact. This latter feature suggests a low green density and little solid state sintering during heating. If the solid-liquid dihedral angle is close to zero, the penetration of the liquid between particles favorably aids rearrangement. Under these conditions, the rate of deformation of the solid-liquid mixture is determined by viscous flow. In Figure 3.18 the rate of deformation is shown for two types of solid-liquid mixtures. A Newtonian system has a deformation rate

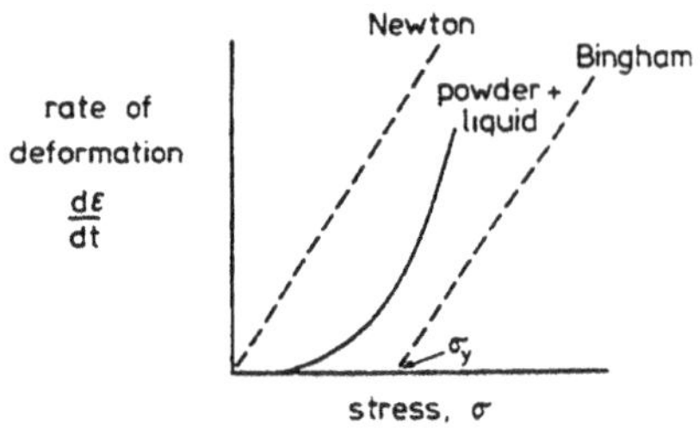

Figure 3.18 The stress-strain plots for two forms of viscous flow. Solid-liquid-vapor compacts are intermediate in behavior during the initial stage of liquid phase sintering.

$d\varepsilon/dt$ proportional to the applied stress σ,

$$d\varepsilon/dt = \sigma/\mu \tag{3.9}$$

where ε is the strain and μ is the viscosity. In contrast, a Bingham solid has an effective yield strength σ_y such that,

$$d\varepsilon/dt = (\sigma - \sigma_y)/\mu. \tag{3.10}$$

A solid-liquid mixture is often found to have a behavior which is intermediate between these two simple models. Figure 3.19 shows this behavior for the W-Cu system where the rearrangement shrinkage varies linearly with the capillary force. These experimental measurements by Huppmann and Riegger (35) on the W-Cu system show rearrangement shrinkage $\Delta L/L_o$ varies with the capillary force F as;

$$\Delta L/L_o = a\,(F - F_o) \tag{3.11}$$

where a is a proportionality constant and F_o represents the resistance to rearrangement.

The relative fluidity K of the solid-liquid suspension decreases with an increasing volume fraction of solid V_S as follows (36):

$$K = 1/(1 + c\,V_S) \tag{3.12}$$

with c being a constant typically near 2.5. The absolute fluidity is dependent on the liquid and temperature. The stress causing rearrangement is the capillary force due to a wetting liquid. The viscosity of the solid-liquid-vapor mixture allows initial densification (pore elimination) by particle rearrangement (37). Initially, the pores aid viscous flow; however, with pore elimination the resistance to rearrangement increases. By analog to Equation (3.12) it would be expected that the densification rate will not be constant during

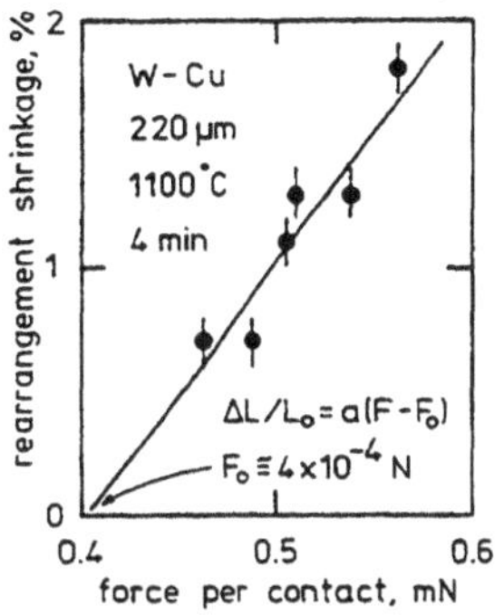

Figure 3.19 The rearrangement shrinkage increases linearly with the capillary force for copper coated tungsten spheres (35).

rearrangement because of a decreasing porosity and concomitant viscosity rise. The rearrangement rate is thought to be fairly independent of the volume fraction of liquid, although the amount of rearrangement depends on the volume fraction of liquid.

G. Solubility

There are two solubilities of concern in liquid phase sintering; the liquid solubility in the solid and the solid solubility in the liquid. Generally, a high liquid solubility in the solid is unfavorable. Such a condition can lead to a transient liquid phase and considerable processing sensitivity. Furthermore, depending on diffusion rates, an unbalanced solubility can lead to Kirkendall porosity and compact swelling during sintering (24,38,39). Alternatively, a high solid solubility in the liquid is favorable. As an illustration, consider the data shown in Figure 3.20 for cemented carbides (40). Both compositions have 18 volume percent liquid; however, the WC-Co system has solid solubility in the liquid, while the WC-Cu system does not. Thus, there is clear evidence that the solubility and the solubility ratio (solid solubility in the liquid divided by liquid solubility in the solid) are important to liquid phase sintering. Note that an increase in sintering temperature for WC-Cu aids densification but sintering temperature does not explain the large difference between the two systems.

The solubility in any system depends on the interactions between the components (4) Solubility is limited by atomic bonding saturation. In liquid phase sintering, the saturation solubility at the sintering temperature determines both the amount and composition of the phases. In turn, the solubility is important to wetting, solution-reprecipitation, grain coarsening, and dimensional changes during sintering. Because solubility depends on pressure (solutions exhibiting negative deviations from ideal behavior have an increasing solubility with pressure), then according to Equation (3.2) there is a particle size effect on solubility. The solubility dependence on spherical particle size is given as follows:

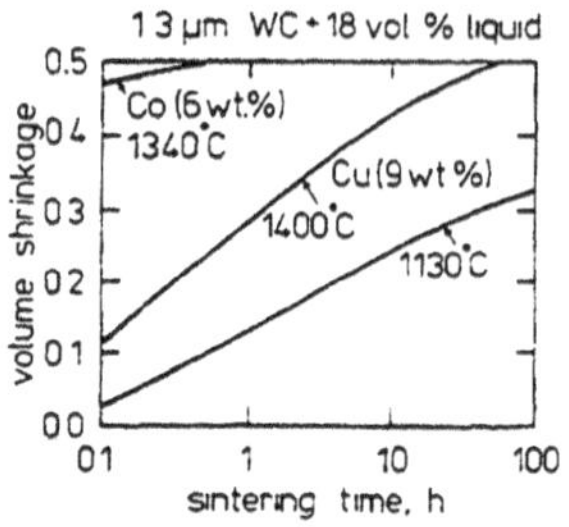

Figure 3.20 The volume shrinkage for liquid phase sintered cemented carbides. The more rapid densification of the WC-Co composition results from the higher solid solubility in the liquid.

$$ln(S/S_o) = 2 \gamma \Omega/(R k T) \tag{3.13}$$

where Ω is the atomic volume, R is the particle radius, S is the solubility of the particle, and S_o is the equilibrium solubility which corresponds to a flat surface.

Small grains will have a higher solubility than large grains. This sets up the condition for grain coarsening by solution-reprecipitation. For example in the iron-copper system the estimated solubility with respect to equilibrium is 1.00048 for an iron particle size of 1 µm (41). Only at very small particle sizes, in the submicron range, is there a measurable solubility change according to Equation (3.13). Thus, particle sizes typical to liquid phase sintering have a small effect on the basic system solubility behavior. However, with prolonged sintering these small solubility effects can have a profound effect on the sintered microstructure.

H. Interdiffusion, Reaction, and Homogenization

A high melting species and a low melting liquid will often react to form an intermediate compound phase and cause swelling. The formation of such a compound can inhibit liquid spreading and subsequent sintering. Since diffusion rates are inversely dependent on the absolute melting temperature (42), the liquid forming additive is expected to show the greater diffusivity in this compound. Thus, concentration profiles in a compact during liquid phase sintering will not be symmetric. Capillarity will cause a wetting liquid to preferentially penetrate along interparticle boundaries (43). Furthermore, depending on reactions, diffusivities, and solubilities, pore formation can occur during heating. Since the reaction energy is high in comparison to the surface energy, pores are most likely to form during a reaction between components. Bugakov (44) found that interdiffusion leading to intermediate phase formation always first gave the compound phase with the largest negative free energy of formation. Soon after forming this compound, thermodynamic equilibrium took over and other lower stability phases formed as would be expected from the equilibrium phase diagram. In cases of a strong reaction between solid and liquid, wetting is good and spreading is rapid.

However, these same conditions often lead to pore formation, swelling, and penetration of grain boundaries by the liquid.

In many systems, interdiffusion is actually reactive in nature. For example, in sintering mixtures of Ti and Al, Savitskii and Burtsev (45) observed spontaneous heating from 660°C to over 1000°C when the aluminum melted. Under such cases of reaction controlled liquid phase sintering, the formation of intermediate phases occurs quickly. In contrast, diffusion controlled formation of an intermediate phase will have a square-root of time dependence. For such cases the rate of homogenization for mixed powders will depend on the parameter χ (46),

$$\chi = D\ t/R^2 \tag{3.14}$$

where D is the controlling diffusivity, t is the isothermal time, and R is the particle radius. The amount of reaction product formed during liquid phase sintering depends on the diffusivity through any intermediate phase between the liquid and solid. The intermediate phase initially thickens with the square-root of time. The kinetics of reaction at a solid-liquid interface depend on both dissolution and reaction steps. If dissolution is the primary process, then an equilibrium sequence of phases will be observed at the interface. However, if reaction is the primary process, then the phase with the lowest free energy of formation will form first and will probably hinder interdiffusion and homogenization because of low diffusivity. For diffusion controlled reactions between mixed powders, Beretka (47,48) has shown the best model as,

$$1 - (1 - \beta)^{1/3} = \Gamma\ t^{1/2} \tag{3.15}$$

where β is the fraction transformed, t is the isothermal time, and Γ is a rate constant dependent on temperature as follows;

$$\Gamma = \Gamma_o\ exp[-Q/(kT)] \tag{3.16}$$

where Q is the activation energy and Γ_o is the frequency factor.

Considering the exponential temperature dependence of the rate constant, slow heating is expected to give greater homogenization. For transient liquid phase sintering this reduces the amount of liquid and greatly affects the sintering rate.

I. Summary

Interfacial energy is important since it establishes the main driving force for liquid phase sintering. The interfacial energy affects both the solubility and capillarity, as a consequence it is important to solution-reprecipitation, coalescence, liquid spreading, and particle rearrangement. Beyond a role as the major driving force during liquid phase sintering, interfacial energies establish important microstructural features like contiguity and grain size. The interfacial energy originates with the fundamental nature of the atomic bonds and is sensitive to contamination and segregation.

Unfortunately, accurate values for interfacial energies are hard to obtain because of easy contamination. From a practical view, actual values for the interfacial energies will be variable during liquid phase sintering because of reactions, contamination, and segregation. However, the basic driving forces and effects on the microstructure as described in this chapter will be applicable in spite of these practical difficulties. In the next four chapters, the basic driving forces will be used to explain in sequence the events observed during liquid phase sintering. This discussion will initially focus on the classic persistent liquid phase sintering process, with subsequent discussion on the special processing options involving transient liquids, reactions, and pressure.

J. References

1. J. Gurland and J. T. Norton, "Role of the Binder Phase in Cemented Tungsten Carbide-Cobalt Alloys," *Trans. AIME*, 1952, vol.194, pp.1051-1056.
2. L. E. Murr, *Interfacial Phenomena in Metals and Alloys*, Addison-Wesley Publ., Reading, MA, 1975.
3. C. Isenberg, *The Science of Soap Films and Soap Bubbles*, Tieto Ltd., Avon, United Kingdom, 1978.
4. J. H. Hildebrand and R. L. Scott, *The Solubility of Nonelectrolytes*, third edition, Dover Publ., New York, NY, 1964.
5. R. S. Burdon, *Surface Tension and the Spreading of Liquids*, second edition, Cambridge University Press, Cambridge, United Kingdom, 1949.
6. V. N. Eremenko, Y. V. Naidich, and I. A. Lavrinenko, *Liquid-Phase Sintering*, Consultants Bureau, New York, NY, 1970.
7. F. V. Lenel and T. Pecanha, "Observations on the Sintering of Compacts from a Mixture of Iron and Copper Powders," *Powder Met.*, 1973, vol.16, pp.351-365.
8. R. M. German, "Origin of the Necklace Structure in Liquid Phase Sintering," *Inter. J. Powder Met.*, 1986, vol.23, in press.
9. A. M. Stoneham, "Ceramic Surfaces: Theoretical Studies," *J. Amer. Ceramic Soc.*, 1981, vol.64, pp.54-60.
10. N. Eustathopoulos and J. C. Joud, "Interfacial Tension and Adsorption in Metallic Systems," *Current Topics in Materials Science*, vol.4, E. Kaldis (ed.), North-Holland Publ., Amsterdam, Netherlands, 1980, pp.281-360.
11. J. J. Burton and E. S. Machlin, "Prediction of Segregation to Alloy Surfaces from Bulk Phase Diagrams," *Phys. Rev. Lett.*, 1976, vol.37, pp.1433-1436.
12. E. D. Hondros and M. P. Seah, "Segregation to Interfaces," *Inter. Metals Revs.*, 1977, vol.22, pp.262-301.
13. D. Camel, N. Eustathopoulos, and P. Desre, "Chemical Adsorption and Temperature Dependence of the Solid-Liquid Interfacial Tension of Metallic Binary Alloys," *Acta Met.*, 1980, vol.28, pp.239-247.
14. M. P. Seah, "Grain Boundary Segregation," *J. Phys. F: Metal Phys.*, 1980, vol.10, pp.1043-1064.
15. J. F. Kuzmick and E. N. Mazza, "Studies on Control of Growth or Shrinkage of Iron-Copper Compacts During Sintering," *Trans. AIME*, 1950, vol.188, pp.1218-1219.
16. C. Durdaller, "The Effect of Additions of Copper, Nickel and Graphite on the Sintered Properties of Iron-Base Sintered P/M Parts," *Prog. Powder Met.*, 1969, vol.25, pp.73-100.

17. I. A. Aksay, C. E. Hoge and J. A. Pask, "Wetting Under Chemical Equilibrium and Nonequilibrium Conditions," *J. Phys. Chem.*, 1974, vol.78, pp.1178-1183.
18. I. A. Aksay, C. E. Hoge, and J. A. Pask, "Phase Distribution in Solid-Liquid-Vapor Systems," *Surfaces and Interfaces of Glass and Ceramics*, V. D. Frechette, W. C. Lacourse and V. L. Burdick (eds.), Plenum Press, New York, NY, 1974, pp.299-321.
19. J. White, "Microstructure and Grain Growth in Ceramics in the Presence of a Liquid Phase," *Sintering and Related Phenomena*, G. C. Kuczynski (ed.), Plenum Press, New York, NY, 1973, pp.81-108.
20. H. Fischmeister, A. Kannappan, L. Ho-Yi, and E. Navara, "Grain Growth During Sintering of W-Cu-Ni Alloys," *Phys. Sintering*, 1969, vol.1, pp.G1-G13.
21. P. E. D. Morgan and M. S. Koutsoutis, "Phase Studies Concerning Sintering in Aluminas Doped with Ti(+4)," *J. Amer. Ceramic Soc.*, 1985, vol.68, pp.C156-C158.
22. D. Y. Kim and A. Accary, "Mechanisms of Grain Growth Inhibition During Sintering of WC-Co Based Hard Metals," *Sintering Processes*, G. C. Kuczynski (ed.), Plenum Press, New York, NY, 1980, pp.235-244.
23. T. J. Whalen and M. Humenik, "Sintering in the Presence of a Liquid Phase," *Sintering and Related Phenomena*, G. C. Kuczynski, N. Hooton and C. Gibbon (eds.), Gordon and Breach, 1967, New York, NY, pp.715-74
24. D. J. Lee and R. M. German, "Sintering Behavior of Iron-Aluminum Powder Mixtures," *Inter. J. Powder Met. Powder Tech.*, 1985, vol.21, pp.9-21.
25. W. D. Kingery, "Densification During Sintering in the Presence of a Liquid Phase. I. Theory," *J. Appl. Phys.*, 1959, vol.30, pp.301-306.
26. R. B. Heady and J. W. Cahn, "An Analysis of the Capillary Forces in Liquid-Phase Sintering of Spherical Particles," *Metall. Trans.*, 1970, vol.1, pp.185-189.
27. H. Emi, S. Endo, C. Kanaoka, and S. Kawai, "Measurement of Forces due to a Liquid Bridge between Spherical Solid Particles," *Int. Chem. Eng.*, 1979, vol.19, pp.300-306.
28. B. Derjaguin, "Concerning the Paper: 'The Effect of Capillary Liquid on the Force of Adhesion between Spherical Solid Particles,'" *J. Colloid Interface Sci.*, 1968, vol.26, p.253.
29. H. M. Princen, "Comments on 'The Effects of Capillary Liquid on the Force of Adhesion between Spherical Solid Particles,'" *J. Colloid Interface Sci.*, 1968, vol.26, pp.249-253.
30. K. S. Hwang, "Analysis of Initial Stage Sintering in the Solid and Liquid Phase," Ph.D. Thesis, Rensselaer Polytechnic Institute, Troy, NY, 1984.
31. V. Smolej and S. Pejovnik, "Some Remarks on the Driving Force for Liquid-Phase Sintering," *Z. Metallkde.*, 1976, vol.67, pp.603-605.
32. W. Pietsch and H. Rumpf, "Haftkraft, Kapillardruck, Flussingkeitsvolumen und Grenzwinkel einer Flussigkeitsbrucke zwischen zwei Kugeln," *Chemie-Ing.-Techn.*, 1967, vol.39, pp.885-893.
33. G. Mason and W. C. Clark, "Liquid Bridges between Spheres," *Chem. Eng. Sci.*, 1965, vol.20, pp.859-866.
34. Y. V. Naidich, I. A. Lavrinenko, and V. Y. Petrishchev, "Study on the Capillary Adhesive Forces Between Solid Particles with a Liquid Layer at the Points of Contact. 1. Spherical Particles," *Soviet Powder Met. Metal Ceram.*, 1965, vol.4, pp.129-133.
35. W. J. Huppmann and R. Riegger, "Modelling of Rearrangement Processes

in Liquid Phase Sintering," *Acta Met.*, 1975, vol.23, pp.965-971.
36. W. D. Kingery, "Sintering in the Presence of a Liquid Phase," *Ceramic Fabrication Processes*, W. D. Kingery (ed.), John Wiley, New York, NY, 1958, pp.131-143.
37. A. Crowson and J. W. Burlingame, "Activated Sintering of Steel Powders," *Processing of Metal and Ceramic Powders*, R. M. German and K. W. Lay (eds.), The Metallurgical Society, Warrendale, PA, 1982, pp.199-211.
38. A. P. Savitskii and N. N. Burtsev, "Compact Growth in Liquid Phase Sintering," *Soviet Powder Met. Metal Ceram.*, 1979, vol.18, pp.96-102.
39. A. P. Savitskii and L. S. Martsunova, "Effect of Solid-State Solubility on the Volume Changes Experienced by Aluminum During Liquid-Phase Sintering," *Soviet Powder Met. Metal Ceram.*, 1977, vol.16, pp.333-337.
40. R. F. Snowball and D. R. Milner, "Densification Processes in the Tungsten Carbide-Cobalt System," *Powder Met.*, 1968, vol.11, pp.23-40.
41. W. D. Kingery "Sintering in the Presence of a Liquid Phase," *Kinetics of High-Temperature Processes*, W. D. Kingery (ed.), John Wiley, New York, NY, 1959, pp.187-194.
42. A. M. Brown and M. F. Ashby, "Correlations for Diffusion Constants," *Acta Met.*, 1980, vol.28, pp.1085-1101.
43. W. Kehl and H. F. Fischmeister, "Liquid Phase Sintering of Al-Cu Compacts," *Powder Met.*, 1980, vol.23, pp.113-119.
44. V. Z. Bugakov, *Diffusion in Metals and Alloys*, National Technical Information Service, Springfield, VA, 1971.
45. A. P. Savitskii and N. N. Burtsev, "Effect of Powder Particle Size on the Growth of Titanium Compacts During Liquid-Phase Sintering with Aluminum," *Soviet Powder Met. Metal Ceram.*, 1981, vol.20, pp.618-621.
46. R. W. Heckel, R. D. Lanam, and R. A. Tanzilli, "Techniques for the Study of Homogenization in Compacts of Blended Powders," *Advanced Experimental Techniques in Powder Metallurgy*, J. S. Hirschhorn and K. H. Roll (eds.), Plenum Press, New York, NY, 1970, pp.139-188.
47. J. Beretka and T. Brown, "Effect of Particle Size on the Kinetics of the Reaction Between Magnesium and Aluminum Oxides," *J. Amer. Ceramic Soc.*, 1983, vol.66, pp.383-388.
48. J. Beretka, "Kinetic Analysis of Solid-State Reactions Between Powdered Reactants," *J. Amer. Ceramic Soc.*, 1984, vol.67, pp.615-620.

CHAPTER FOUR

Initial Stage Processes: Solubility and Rearrangement

A. Overview

The classic treatment of persistent liquid phase sintering breaks the process into three stages. These three stages overlap, with the initial stage being the shortest. The initial stage is often referred to as the rearrangement stage because of the significant densification resulting from particle repacking concurrent with liquid formation. Additionally, considerable densification or swelling is apparent during heating to the liquid formation temperature. In spite of the short duration, the changes induced in the initial stage are often the most significant. Cannon and Lenel (1) were the first to formally study rearrangement. Subsequently, Kingery (2) gave a mathematical treatment of the primary factors. Eremenko et al. (3) have refined and updated these treatments and established the basic theory as being correct.

Dramatic microstructural changes occur during the initial stage. The events include solid state sintering, coalescence, liquid spreading, capillary attraction, particle sliding, diffusional homogenization, and particle disintegration and fragmentation (4). The liquid formation and associated capillary force cause rearrangement and clustering in the microstructure. Rearrangement occurs both between particles and between clusters of particles.

Early discussions on liquid phase sintering applied the heavy alloy mechanism to all systems (5). This classical process of liquid phase sintering has a focus on rearrangement as a densification step. According to the heavy alloy mechanism there are five criteria necessary for significant rearrangement on melt formation; i) solubility of the solid in the liquid, ii) low contact angle, iii) small dihedral angle, iv) low degree of solid state interparticle bonding, and v) loose powder structure. Swelling becomes more probable as fewer of these criteria are satisfied. Table 4.1 gives a comparison of five systems processed by liquid phase sintering (6). It shows three basic material characteristics necessary for rearrangement; contact angle, dihedral angle, and solid solubility in the liquid. This table demonstrates the association between rearrangement densification and the first three criteria listed above. The last two criteria (iv and v) depend on the processing conditions and are not material parameters. Caution is necessary when comparing systems because of the significant effects from processing factors like particle size, heating rate, and green density. As an example, consider the shrinkage data for WC-Co shown in Figure 4.1. The shrinkage is shown versus the

TABLE 4.1
Initial Stage Rearrangement

System	Contact Angle, degree	Dihedral Angle, degree	Solubility at.%	Rearrangement
Al_2O_3-Ni	138	---	0	liquid exudes
Fe-Ag	70	158	0	none
TiC-Ni	30	45	10	extensive
Fe-Cu	0	27	5	extensive
TiC-Mo-Ni	0	0	10	extensive

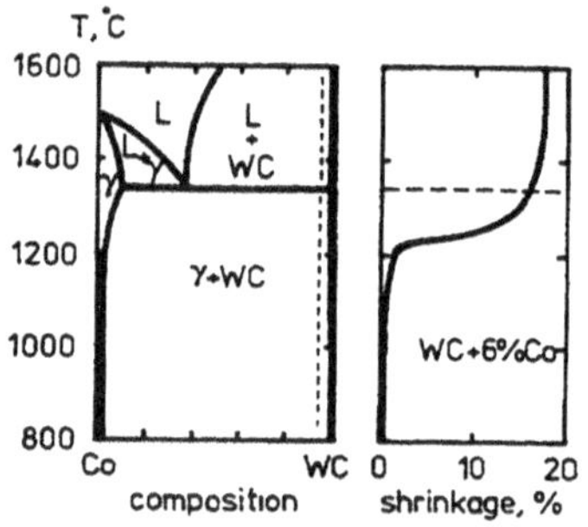

Figure 4.1 The WC-Co binary phase diagram is shown on the left and the sintering shrinkage versus temperature is shown on the right for a 6% Co alloy (3).

sintering temperature in parallel with the WC-Co binary phase diagram. Considerable densification occurs during heating below the eutectic temperature (7), eliminating rearrangement. In more complex systems, like porcelain, reactions during sintering continuously change the viscosity, interfacial energy, and amount of liquid. Thus, in many cases the classic models are of limited value since they can only indicate general factors. In spite of such shortcomings, the general criteria for liquid phase sintering are an excellent starting point for identification of successful systems.

Both the processing conditions and material characteristics influence the behavior of a liquid phase sintering material. Table 4.2 contrasts some of the parameters associated with liquid phase sintering and their general effects on swelling or shrinkage during the first few minutes of sintering. Shrinkage is favored by systems satisfying the heavy alloy criteria. Alternatively, swelling is possible during heating to the liquid formation temperature if the diffusion

TABLE 4.2
Characteristics Affecting Initial Stage Behavior

Factor	Swelling	Shrinkage
solid solubility in liquid	low	high
liquid solubility in solid	high	low
diffusivities	unequal	equal
additive particle size	large	small
base particle size	large	small
green density	high	low
contact angle	high	low
dihedral angle	high	low
temperature	low	high
time	short	long

rates or solubilities are unbalanced. Another problem is porosity due to intermediate compound formation between the mixed components. These various factors inhibit a straightforward treatment of initial stage events. Consequently, this chapter will separate the main factors to establish general behavior patterns. A few specific systems will be described in detail to illustrate how the basic concepts can be combined.

B. Solubility Effects

Prior to liquid formation several important events take place including solid state sintering. In this chapter, the focus is on the processes occurring during heating and shortly after melt formation. The possible events upon melting are rapid shrinkage due to particle rearrangement, expansion due to liquid penetration of grain boundaries, dissolution of surface irregularities, and formation of new contacts. The relative importance of these events depends on the material, heating rate, atmosphere, particle size, and green density.

A high solid solubility in the liquid aids densification by rearrangement (8-11). Because solubility affects all stages of liquid phase sintering, model studies are often performed on insoluble systems such as W-Ag and W-Cu to isolate the rearrangement processes.

An unbalanced diffusivity between the additive (minor component which forms the liquid) and base (major component) effects the dimensions during heating. The rate of interchange between the additive and base can be given by a diffusivity ratio D_R,

$$D_R = D_B/D_A \tag{4.1}$$

where D stands for the solute diffusivity and the subscripts A and B denote the additive and base, respectively. These diffusivities correspond to the individual diffusion rates for the base in the additive and vice versa. If the diffusivity ratio is larger than 1, then preferential diffusion of the base into the additive is expected. Alternatively, a diffusivity ratio less than 1 favors preferential additive flow into the base. In both cases, Kirkendall porosity is expected as the diffusivities differ from one another (12). After sufficient time, the diffusivity ratio becomes less important since equilibrium effects then dominate. Thus, unbalanced diffusivity ratios contribute to initial pore formation; however, sustained swelling or shrinkage depends on the solubility ratio.

As the solubility of the solid in the liquid increases, processes other than rearrangement contribute to initial densification. Limited densification occurs with an insoluble system since rearrangement is the only active process. Extensive densification is observed with a high solid solubility in the liquid. In contrast, a low solid solubility in the liquid coupled to a high solubility of the liquid in the solid gives swelling. The solubility ratio S_R is defined as,

$$S_R = S_B/S_A \tag{4.2}$$

where S represents the solubility and the subscript denotes B for base (major phase) solubility in the additive and A for additive (minor phase) solubility in the base. Figure 4.2 shows a schematic plot of the solubility effect. A low solubility ratio leads to pore formation at the prior additive particle site, while a high solubility ratio gives densification. The role of the solid solubility in the liquid is related to lubrication, sliding and surface smoothing. Another example of the solubility role in initial densification is shown in Figure 4.3. This plot shows the shrinkage of chromium carbide in a nickel-copper liquid at 1350°C. As the nickel content increases, the carbide solubility in the liquid increases. The higher solubility gives faster densification as well as greater densification. Additionally, note the rapid initial densification in the first ten minutes of sintering, which corresponds to the initial stage.

For systems with a high solid solubility in the additive, extensive densification is observed prior to formation of the first liquid. As an example, in WC-Co with over 8% Co, no liquid phase is needed to obtain full density (13). At lower Co contents (like 5%), a few minutes in the liquid range is necessary for full densification.

In the balance of this section the solubility effects on initial shrinkage and swelling prior to rearrangement will be detailed. In these cases, mass transfer occurs during heating. Dimensional changes accompany the mass transfer and are dependent on the solubility and diffusivity ratios. After liquid formation, other factors associated with particle repacking will control densification.

Savitskii et al. (14) have shown the importance of the initial porosity and the volumetric solubility in the case of a high solubility ratio. The porosity ε after dissolution of the base into the additive (assuming no significant quantity of additive dissolves into the base) depends on the initial porosity ε_o as follows:

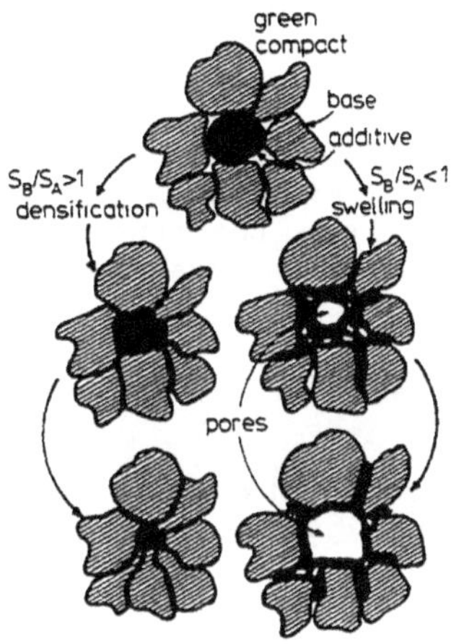

Figure 4.2 A schematic diagram contrasting the effects of solubility on densification or swelling during sintering.

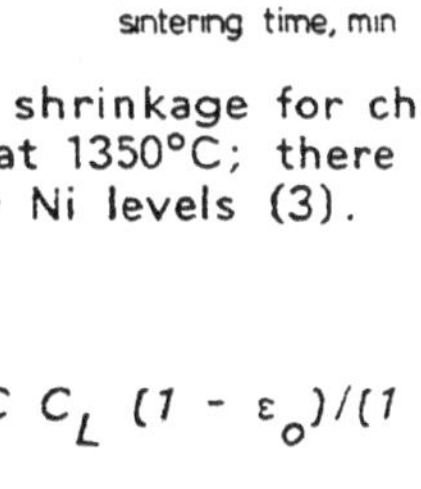

Figure 4.3 Solubility effect on shrinkage for chromium-carbide in 50 vol.% liquid Ni-Cu melts at 1350°C; there is increased solubility for the carbide with higher Ni levels (3).

$$\varepsilon = \varepsilon_o - C\, C_L\, (1 - \varepsilon_o)/(1 - C - C_L) \tag{4.3}$$

which can also be written as follows:

$$\varepsilon = [\varepsilon_o\, (1 - C)(1 - C_L) - C\, C_L]/(1 - C - C_L) \tag{4.4}$$

where C is the concentration (volume fraction) of liquid forming additive and the concentration (volume fraction) of solid dissolved into the additive (or liquid) is given as C_L.

A consequence of the above volume conservation equation is that the porosity just after forming a liquid varies linearly with the initial porosity. Increases in the volume fraction of additive will decrease the porosity after liquid formation. Also the greater the solubility of the solid in the liquid, the greater the densification. A small amount of liquid forming additive is sufficient when the base has a high solubility in the liquid. For full densification, a final porosity of zero is desired. The minimum volume fraction of liquid forming additive needed for maximum densification by dissolution events only is given as C_m, where

$$C_m = \varepsilon_o (1 - C_L)/(C_L - C_L\varepsilon_o + \varepsilon_o). \tag{4.5}$$

An example of the linear relation between the initial (green) porosity and the final porosity is illustrated in Figure 4.4 for Al-1% Cu sintered at 640°C. If there were no interaction between the powders, then the final porosity would equal the initial porosity. Alternatively, the linear behavior predicted by Equation (4.4) is readily evident.

In most binary systems, the liquidus moves closer to the solid composition as temperature increases. Thus, according to Equation (4.4) higher temperatures are favorable for liquid phase sintering because of increased solid solubility in the liquid and a greater quantity of liquid.

Mixed powders can undergo an exothermic reaction during heating. For such systems, it is typical for the reaction energy to greatly exceed the surface energy. As a consequence pore formation occurs in spite of the concomitant creation of new surfaces. Up to now a low additive solubility in the base (solid) has been assumed. However, this solubility can also greatly effect the dimensional change in sintering. Additives with a high solubility in the solid will dissolve during heating, forming a pore at the prior additive site. Figure 4.5 shows the site of a titanium particle in an iron matrix after heating through the eutectic temperature of 1085°C. The eutectic liquid has spread and left a pore at the prior titanium particle site. In such cases swelling is observed and rearrangement is slight. This behavior is characteristic of transient liquid phase sintering.

The problem of swelling during melt formation has been treated by Savitskii et al. (14-19). For a material combination where the solubility ratio is small, the porosity will vary with the concentration of additive C and the fraction which has reacted f as follows:

$$\varepsilon = \varepsilon_o + f\,C\,(1 - \varepsilon_o) \tag{4.6}$$

which can be rewritten as follows:

$$\varepsilon = f\,C + \varepsilon_o\,(1 - f\,C) \tag{4.7}$$

which shows the effect of an increasing initial porosity or concentration of additive on the final porosity. Examples of swelling behavior are shown in Figures 4.6 and 4.7 for the Al-Zn system. The porosity after sintering is shown versus the initial (green) porosity. The linear form predicted by

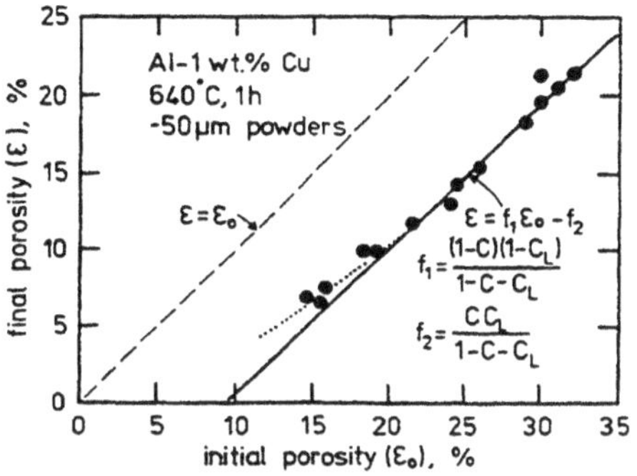

Figure 4.4 The final porosity for Al-Cu compacts shown versus the initial porosity (14). The densification behavior follows the form predicted by Equation (4.4).

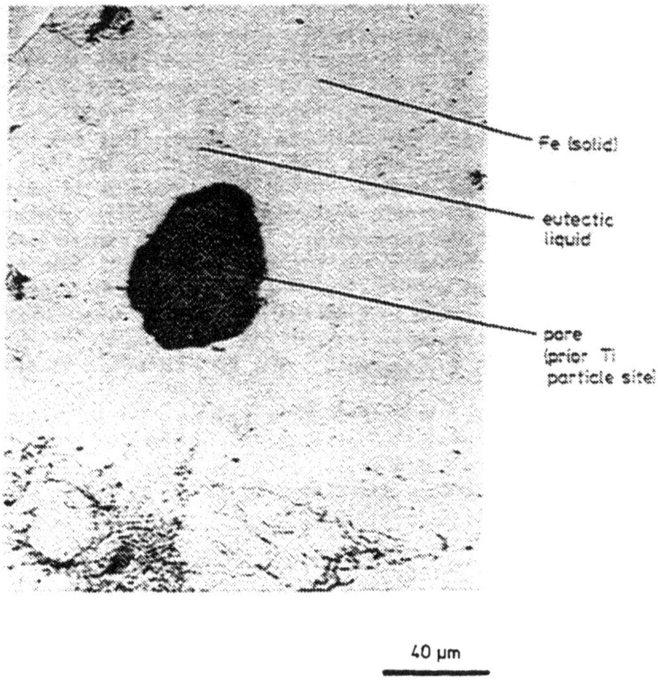

Figure 4.5 Pore formation at the prior additive particle site in Fe-Ti sintered above the 1085°C eutectic temperature (photo courtesy of J. W. Dunlap).

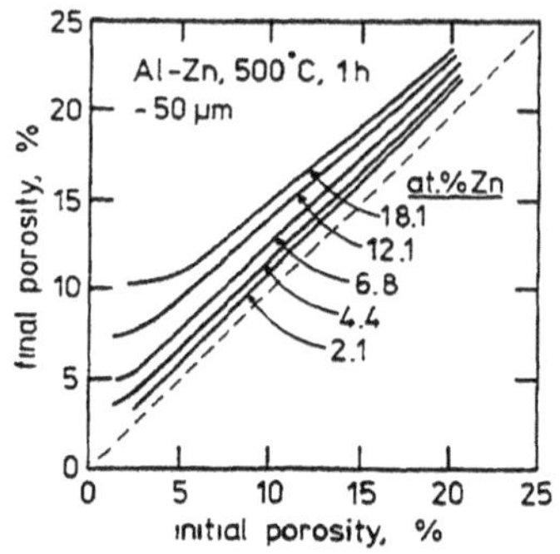

Figure 4.6 The final porosity of Al-Zn compacts versus the initial porosity, after sintering at 500°C with various Zn contents (17).

Equation (4.7) is apparent except at the low initial porosities. As the concentration of Zn increases, the final porosity also increases. Equations (4.3) through (4.7) must be modified to account for solid state sintering, rearrangement, intersolubility, stresses due to reactions, and differential diffusion rates (19). This is illustrated in Figure 4.7 for the Al-Zn system, showing the linear behavior of Equation (4.7) holds, but that other factors like particle size must be considered. Figure 4.8 plots the data from Figure 4.7 to show a log-log relation between final porosity and Al particle size at a fixed 20% initial porosity. A larger aluminum particle size gives greater swelling at a constant zinc concentration. In spite of limited prediction capabilities, these models demonstrate the major effect of density and solubility on dimensional change during heating mixed powder systems. There are two other forms of each equation worth noting. For a small solubility ratio system, the densification parameter ψ (change in porosity divided by the change necessary to obtain full density) can be expressed as follows:

$$\psi = -f\,C\,(1 - \varepsilon_o)/\varepsilon_o \tag{4.8}$$

where a negative value indicates swelling. In terms of linear dimensional change $\Delta L/L_o$,

$$\Delta L/L_o = 1 - (1 - f\,C)^{-1/3} \tag{4.9}$$

where a negative value for the dimensional change indicates swelling. For systems with a large solubility ratio, the equivalent forms follow:

$$\psi = C\,C_L(1 - \varepsilon_o)/[\varepsilon_o(1 - C - C_L)] \tag{4.10}$$

$$\Delta L/L_o = 1 - [(1 - C - C_L)/(1 - C)(1 - C_L)]^{1/3}. \tag{4.11}$$

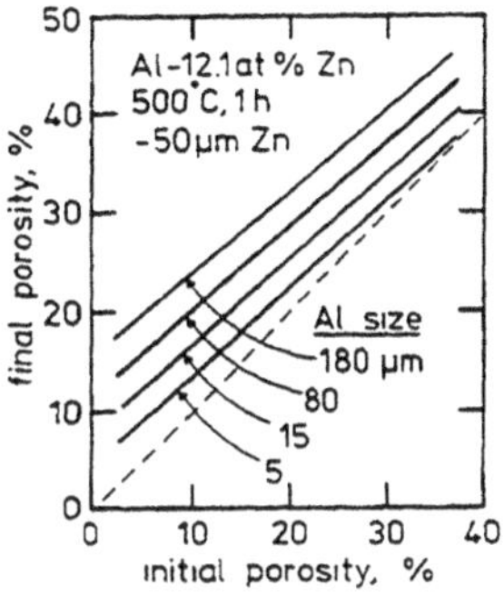

Figure 4.7 The effect of Al particle size on the swelling of Al-12.1% Zn compacts (17).

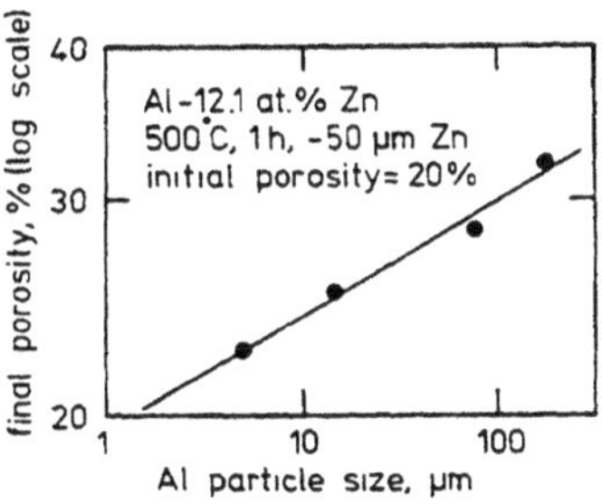

Figure 4.8 The final porosity of Al-12.1% Zn compacts versus the Al particle size (log-log plot), showing the increased swelling with particle size.

For more complex systems, such as those with intermediate compounds or mutual solubility, Savitskii (17) shows various modifications of Equations (4.4) and (4.7). Thus, the dimensional changes during heating can be categorized in terms of solubility and rearrangement events. Table 4.3 compares the behavior found for several binary metallic systems to that predicted from the solubility ratio. The solubility ratio provides an accurate prediction of basic initial swelling or shrinkage tendencies. The actual amount of dimensional change depends on processing conditions.

The use of prealloyed powders provides one alternative to controlling mass flow during heating. Additionally, long sintering times can offset the effects of swelling reactions. Shrinkage can be expected when the volume of liquid increases with time. Alternatively, swelling occurs when the diffusivity ratio differs substantially from unity, the volume of liquid decreases with time (because of a high additive solubility in the solid), or when an intermediate

TABLE 4.3

Solubility Effects on Densification

Base B	Additive A	Solubility Ratio, (at. %)	Behavior	Reference
Al	Zn	0.004	swell	17
Cu	Al	0.1	swell	20
Cu	Sn	0.001	swell	21
Cu	Ti	4	shrink	22
Fe	Al	0.02	swell	23
Fe	B	7	shrink	24
Fe	Cu	0.07	swell	25
Fe	Sn	0.01	swell	26
Fe	Ti	3	shrink	27
Mo	Ni	20	shrink	28
Ti	Al	0.0003	swell	29
W	Fe	5	shrink	30

compound forms between the constituents (31,32). It is observed that the amount of swelling can be reduced for systems with a small solubility ratio by increasing the additive homogeneity. This can be achieved by the use of smaller additive particle sizes or through use of an additive coating on the base powder. In general, the more homogeneous the additive distribution before liquid formation, the more rapid the densification during liquid phase sintering. For this reason, large additive particle sizes are detrimental. Accordingly, it is common to mill the powders together to increase homogeneity, reduce particle sizes, and to break apart agglomerates (33).

C. Melt Formation

After a heating period where diffusivity and solubility effects dominate, melt formation occurs. At this point, the most useful characteristic of the liquid is wetting of the solid. A wetting liquid will spread by capillarity. The major portion of densification by rearrangement occurs within three minutes after melt formation (4,34,35). The shrinkage rate $d(\Delta L/L_o)/dt$ is estimated as follows:

$$d(\Delta L/L_o)/dt = \Delta P\ W/(2\ R\ \mu) \tag{4.12}$$

where ΔP is the capillary pressure, R is the grain radius, W is the liquid thickness, and μ is the viscosity of the liquid. Usually, the measured rates of shrinkage are lower than that predicted by Equation (4.12), indicating that liquid spreading and penetration throughout the microstructure are rate limiting steps. In many systems of practical importance, chemical additions are used to improve wetting or solubility. As an example Ta(C,N) with a nickel liquid phase exhibits poor wetting. However, the addition of vanadium-carbide to the mixture aids densification during liquid phase sintering (36). Alternatively, for the case of poor wetting there will be exuding of the liquid from the material. Figure 4.9 is a photograph of such an exuded liquid after sintering. The solid structure remains intact and porous, while the liquid forms a ball on the surface. Thus, wetting is a significant factor in analyzing initial stage processes.

Beyond wetting, initial stage rearrangement is aided by intersolubility between the solid and liquid (37). Consider the density versus time measurements for WC-Co and WC-Cu shown in Figure 3.20. Cobalt as the liquid gives faster initial stage sintering and results in a substantially higher final density. Alternatively, copper has no solubility for tungsten carbide; thus, it provides slower densification and a lower final density. As the solid solubility increases in the liquid, then particle smoothing occurs on melt formation, making for easier and faster rearrangement. From such considerations it is evident that the initial solid-liquid interaction has considerable influence on sintering behavior. Both liquid wetting and solubility for the solid are favorable attributes, giving densification soon after melt formation.

D. Penetration and Fragmentation

Penetration refers to the liquid flow through the pore and grain structure by a combination of reaction and capillarity. A byproduct of penetration is fragmentation of the solid particles. Microstructural examinations of the events after liquid formation show penetration occurs first, followed by fragmentation (38). In most systems, the liquid initially is not at an equilibrium composition; thus, surface energy is not at equilibrium. Additionally, the initial liquid has a high curvature which aids solubility and penetration. As a consequence, the spreading liquid can penetrate the solid-solid interfaces beyond the equilibrium value corresponding to Equation (2.17) (39). Penetration happens within a few minutes after liquid formation. The actual rate depends on the liquid reactivity, viscosity, and contact angle.

Liquid penetration of the microstructure correlates with the initial dimensional changes. The rate of penetration can be estimated by a formula given by Pejovnik et al. (34),

$$x^2 = r \gamma t \cos\theta / (2 \mu) \tag{4.13}$$

where x is the depth of liquid penetration (distance), r is the pore capillary radius, γ is the liquid-vapor surface energy, θ is the contact angle, t is the time, and μ is the viscosity. According to Equation (4.13), small contact angles increase the penetration rate.

After liquid flow, the liquid shape approaches that described by Beere (40) as illustrated in Figures 2.13 and 2.14. A low dihedral angle is needed for a liquid to remain connected along grain edges. Experiments on liquid infiltration of polycrystalline solids by Jurewicz and Watson (41) show no penetration of grain boundaries even under 8000 atmospheres pressure for a dihedral angle over 60°. This finding is in agreement with the prediction of Beere.

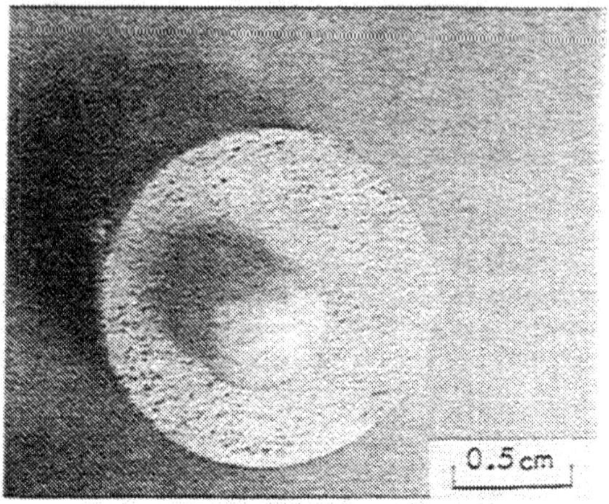

Figure 4.9 Exuded liquid due to poor wetting. The exuded liquid has formed a ball on the top surface of the compact during sintering (photo courtesy of L. Shaw).

Several investigations have reported disintegration and fragmentation of polycrystalline particles by a liquid (10,34,38,42,43). Penetration can fully disintegrate the solid grain structure if the dihedral angle is zero. This results in a dispersed grain size smaller than the initial particle size. A schematic diagram of the fragmentation process is shown in Figure 4.10. The initial solid particles are either polycrystalline or aggregates. The particles undergo rearrangement on melt formation. Subsequently, the liquid penetrates the grain boundaries and gives fragmentation. Fragmentation is usually accompanied by swelling as the grains are separated by the penetrating liquid. Subsequent densification occurs as repacking and solution-reprecipitation begin. Fragmentation requires a solubility of the solid in the liquid.

Disintegration of solid interfaces formed during heating can aid densification. However, alloying additives or impurities can affect grain boundary penetration. A classic example is Fe-Cu-C where the carbon inhibits liquid copper penetration of iron grain boundaries (44). Carbon lowers the solid-solid grain boundary energy, giving a larger dihedral angle and less liquid penetration. From a practical view, swelling due to melt penetration of grain boundaries can be minimized by using porous particles (capillary sites within the particles), lower green densities, and alloying.

E. Contact Force

The contact points between wetted particles are under a strong compressive force (45-49). On liquid formation the contact angle, particle size, and amount of liquid determine the contact forces, as discussed in Chapter 3. For a particle size of one micrometer and a nominal surface energy of one joule per meter square, a force as large as 100 atmospheres is exerted on contacting spheres. During the initial stage of liquid phase

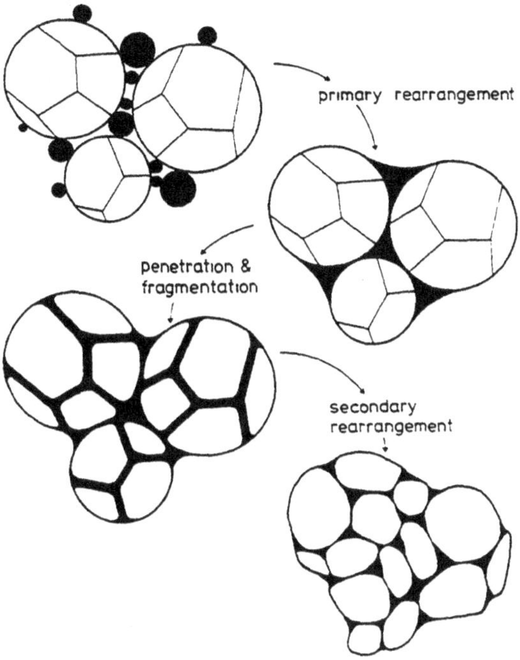

Figure 4.10 A schematic diagram of rearrangement and fragmentation of polycrystalline particles.

sintering, the rearrangement of particles or fragments depends on a large attractive force. Only for small contact angles and small particle sizes is there an appreciable attractive force. At large contact angles, repulsive forces exist and swelling can occur on melt formation.

Accompanying rearrangement there is a pore size change. Typically, the large pores grow while the small pores shrink. Thus, regions with higher packing densities densify preferentially. At large contact angles, repulsive forces exist and the amount of swelling is proportional to the amount of liquid. Figure 4.11 shows this effect for W-Cu where the tungsten has been treated to vary the contact angle (50). A low contact angle gives shrinkage for all liquid contents, while a high contact angle gives swelling at low liquid contents. The relation between shrinkage, swelling, contact angle, and volume fraction of liquid is shown in Figure 4.12. For reference, three contact situations are diagrammed next to the plot. For a dihedral angle greater than zero, there will be a stable neck between the particles which can inhibit rearrangement. At large contact angles and large quantities of liquid, swelling is anticipated. As the particle size increases, the amount of liquid for optimal rearrangement increases (51).

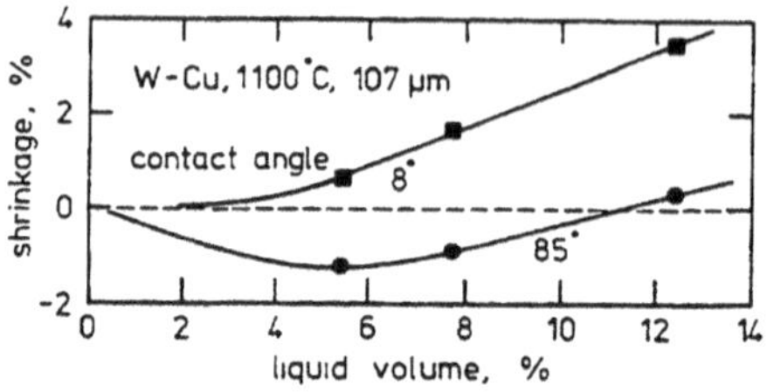

Figure 4.11 The effect of contact angle on the shrinkage or swelling behavior of W-Cu compacts as a function of the liquid volume (50).

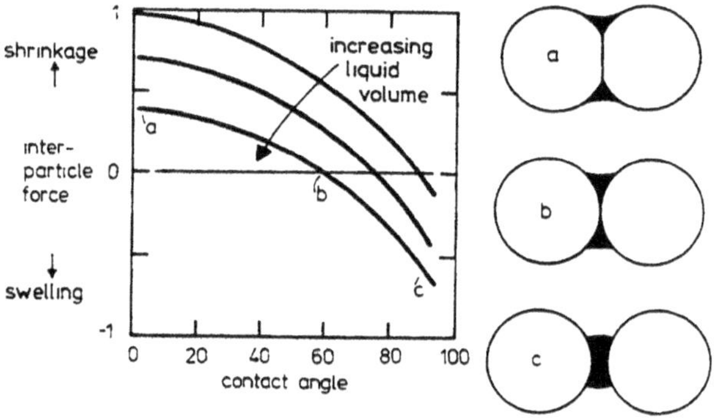

Figure 4.12 The interparticle force variation with the contact angle. The effect of an increasing liquid content is to reduce the force. Three different wetting situations are shown next to the plot.

The situation for irregular particles (typical of ceramics) differs from that for spherical particles. For a low volume fraction of liquid, where neighboring contacts do not merge, the misalignment of the center of gravity due to an irregular shape gives a torque (52). The torque leads to rapid particle rearrangement, bringing flat surfaces into contact. The force on an irregular particle varies with the volume fraction of liquid to the one-third power. Alternatively, the torque varies with the volume fraction of liquid to the one-half power. This gives a major difference in rearrangement behavior between spherical and irregular particles. For a sphere, the force decreases as the volume fraction of liquid increases. An inverse relation holds for irregular particles. Figure 4.13 gives a schematic representation of the difference in the rearrangement force between spherical and irregular particles. In spite of the different force dependencies on the volume fraction of liquid, in practice both systems show better rearrangement at intermediate quantities of liquid where there is less sliding friction. A small particle size and a low contact angle are beneficial to rearrangement, independent of the particle shape.

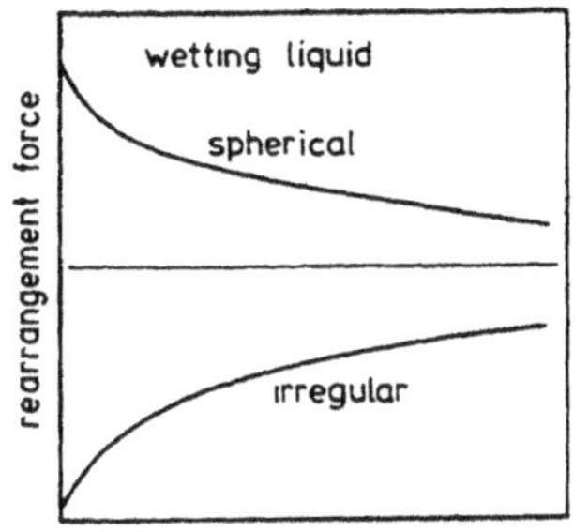

Figure 4.13 A contrast between the rearrangement force for spherical and irregular particles as functions of the liquid content.

The discussion to now has focused on the lower volume fractions of liquid. As the volume fraction of liquid increases, the pores become filled with liquid during the initial stage. The elimination of pores also eliminates the surface energy which drives the rearrangement process. When this occurs the forces contributing to rearrangement diminish to zero.

F. Rearrangement

A wetting liquid creates an attractive force between particles. As a consequence, in a three dimensional network of solid, liquid and vapor, a hydrostatic pressure exists on the pores. Under conditions of low interparticle friction, the particles will repack to a higher coordination. The corresponding density increase is aided by a small particle size and low interparticle friction (smooth particles). Solubility of the solid in the liquid further aids rearrangement because of particle smoothing concurrent with capillary attraction.

Rearrangement is often composed of two stages. Primary rearrangement involves the individual particles. Secondary rearrangement can involve either fragments or clusters of primary particles. The random packing of mixed powders and the uneven distribution of the liquid forming ingredient will produce a process of successive clustering. Figure 4.14 outlines such a process. The liquid forms at the additive particle site. With an excess of liquid, it spreads and creates a cluster of wetted particles with closer packing. The clusters continue to repack as they increase in mass, and accordingly rearrangement becomes slower (4,9,34,53).

Secondary rearrangement can also involve particle disintegration and subsequent repacking of fragments. Liquid attack of the grain boundaries of the primary particles is required for this process. Although primary rearrangement occurs very rapidly, secondary processes can persist up to ten minutes after liquid formation. The rearrangement kinetics of disintegrated particles depends on the fragment size. Repacking of fragments or clusters releases liquid to the intercluster regions. As a consequence, the extent of the microstructure undergoing rearrangement increases with time, while the

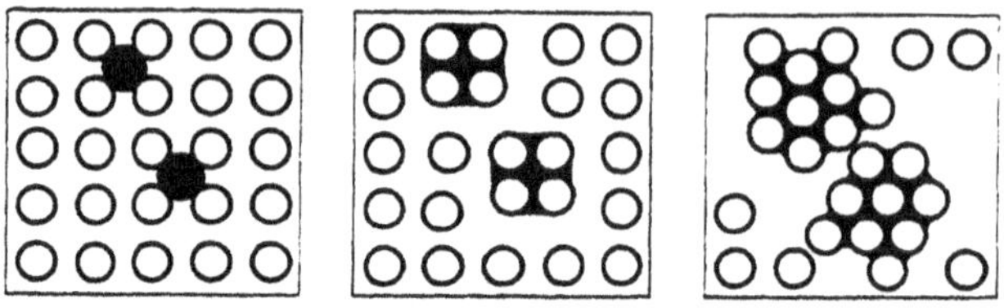

Figure 4.14 Clustering occurs in liquid phase sintering as a consequence of local melt formation and spreading through the microstructure. Eventually, the clusters repack to a higher density.

cluster size continually increases. An example of the importance of secondary rearrangement is given in Figure 4.15. These data show alumina sintering in liquid anorthite (calcia-alumina-silica). The addition of 0.1% magnesia inhibits anorthite penetration of the alumina grain boundaries and prevents secondary rearrangement. As a consequence, densification is retarded by the magnesia addition.

Time is not a significant factor in rearrangement. The three steps are melting, spreading, and penetration. The latter is the slowest. Kingery (2) gave a shrinkage dependence on time to the $1+y$ power,

$$\Delta L/L_o \sim R^{-1}\, t^{1+y} \tag{4.14}$$

where R is the solid particle radius, and $1+y$ is slightly larger than unity. Equation (4.14) assumes the strain associated with viscous flow densification is directly proportional to time and that the surface energy remains constant. Actually, the porosity and capillary pore size continually decrease during rearrangement; thus, and exponent of $1+y$ is proposed to correct for changes in viscosity and capillary force. Experiments have proven Equation (4.14) substantially correct (54-57). Fortes (57) suggests the time exponent is dependent on the volume fraction of liquid and particle mass. Unfortunately, such models are difficult to test accurately. For example, in ordered arrays, the total contraction time is approximately 0.01 s after liquid formation. Random packings are slower because of successive cluster formation with an increasing mass. Various studies have set the value of the exponent $1+y$ at 1.1 to 1.6. Probably a value of 1.3 is a good approximation for most systems undergoing rearrangement.

The volume fraction of liquid is important to the rearrangement stage. At a high volume fraction of liquid, complete densification is possible by rearrangement and pore filling on liquid formation. As the volume fraction of liquid decreases, then other processes like solution-reprecipitation must be active for full densification. The estimated volume shrinkage effect is shown in Figure 4.16. At approximately 30 to 35 volume percent liquid, full densification is expected during rearrangement. At lower volume fractions of liquid, less rearrangement is anticipated and full density requires other events. For an irregular particle shape, the interparticle friction is greater; thus, less

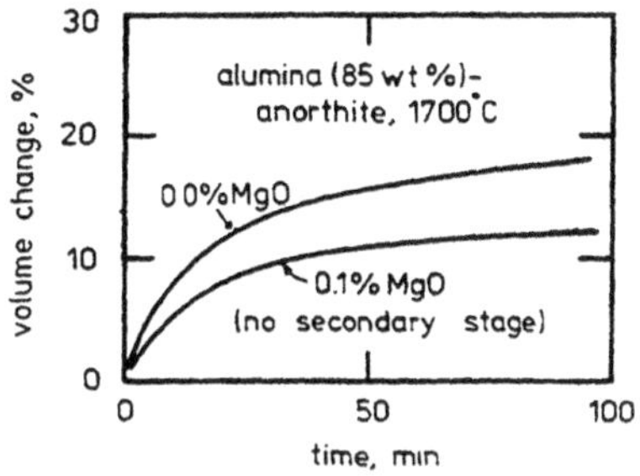

Figure 4.15 The volume change in initial liquid phase sintering for two situations. The addition of magnesia to the alumina-anorthite mixture inhibits secondary rearrangement and reduces densification (70).

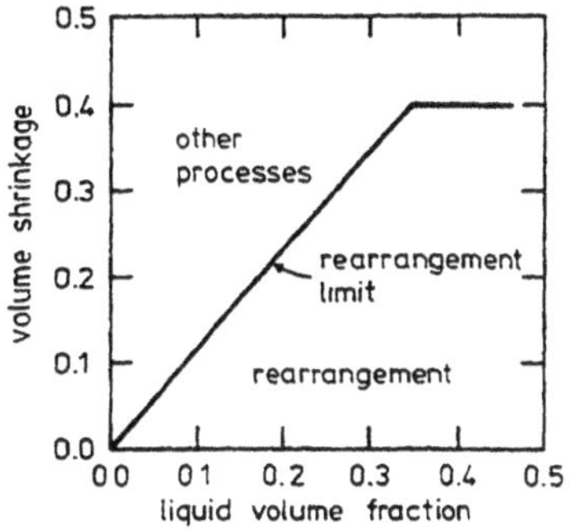

Figure 4.16 The volume shrinkage versus the liquid volume fraction, with the limit possible through rearrangement processes indicated.

densification by rearrangement is observed for a given liquid content. In general, as the amount of liquid increases, the rate of rearrangement is slower while the amount of rearrangement is increased. This is explained by the combined effect of viscosity and capillary force; both the compact viscosity and interparticle capillary force decrease with an increasing volume of liquid. Initially, the decreasing viscosity has a dominant effect (13). However, eventually with large quantities of liquid, viscosity is little changed but the capillary force is decreased. For insoluble systems like WC-Cu, W-Cu, and W-Ag, Eremenko et al. (3) demonstrate maximum densification at approximately 50 volume percent liquid. In many practical liquid phase sintering systems, liquid quantities are kept below 20 volume percent to avoid shape distortion during sintering. At these lower liquid concentrations, shrinkage increases with the volume of liquid.

Cannon and Lenel (1) made early observations on the particle size effect on rearrangement. A fine particle size is beneficial. Figure 4.17 demonstrates the particle size factor using data from the Fe-Cu system. No rearrangement is apparent for the coarser (33 μm) particle size, while for the finest size (3 μm) most of the densification is by rearrangement. Accordingly, Figure 4.8 shows improved densification (less swelling) in the Al-Zn system as the aluminum particle size decreases, indicating improved densification after initial swelling. From capillary observations, the force between particles varies with the inverse of the particle size (58). Thus, the expectation is an inverse relation between rearrangement shrinkage and particle size as given in Equation (4.14). Figure 4.18 illustrates the particle size effect on the densification rate during rearrangement. The data are from Kingery and Narasimhan (54) for various iron particle sizes sintered with a copper-rich liquid. Their finding agrees with the prediction of Equation (4.14). However, the data for diamond in a liquid Cu-Ag-Ti alloy (58) shown in Figure 4.19 gives a different particle size dependence. In this instance the rearrangement shrinkage varied with the particle size to the -1/4 power. Another study failed to find a particle size effect for the insoluble W-Cu system (59). Generally, experiments during rearrangement are not sufficiently accurate to isolate the exact particle size effect. At this time it is unclear what conditions cause the reported differences noted above.

Beyond particle size, particle shape influences rearrangement through the interparticle force. Irregular particles have a greater friction force and will undergo less rearrangement. Thus, little rearrangement is observed with irregular powders like WC at low volume fractions of liquid (53). At high volume fractions of liquid, rearrangement of irregular particles is possible, but occurs in a discontinuous nature.

A low dihedral angle is beneficial to rearrangement (37). A low dihedral angle produces faster and greater rearrangement densification. With a low dihedral angle there is more liquid penetration between particles on melt formation; thus, there is greater opportunity for rearrangement. However, at high packing densities, this same melt penetration process will produce swelling due to particle separation.

Likewise the contact angle has an influence on rearrangement. As seen in the capillary force equation (Equation (3.8)), the better a liquid wets the solid, the greater the force and hence the amount of rearrangement. At large contact angles, no rearrangement is possible. Figure 4.11 shows the data of Petzow and Huppmann (50) for W-Cu rearrangement shrinkage versus volume fraction of liquid with two contact angles. The lower contact angle provides greater rearrangement in this insoluble system.

Green density is another important parameter. As the green density increases, there is more mechanical interlocking and less vapor phase. As a consequence the capillary force responsible for rearrangement is reduced. For insoluble systems, where densification is totally by rearrangement, the final density decreases as the green density decreases. While at high green densities there is greater interparticle friction and less rearrangement. The complex effect of green density is not well established, but in general the densification rate is reduced by a high green density. However, in many cases the final density improves as the green density is increased because of the lower initial porosity. In commercial applications for liquid phase

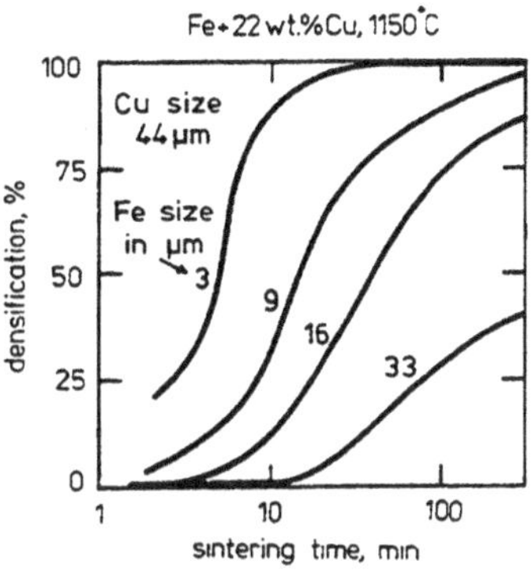

Figure 4.17 Densification in the initial stage of liquid phase sintering for Fe-Cu compacts with various particle sizes of iron; a large iron particle size eliminates rearrangement (1).

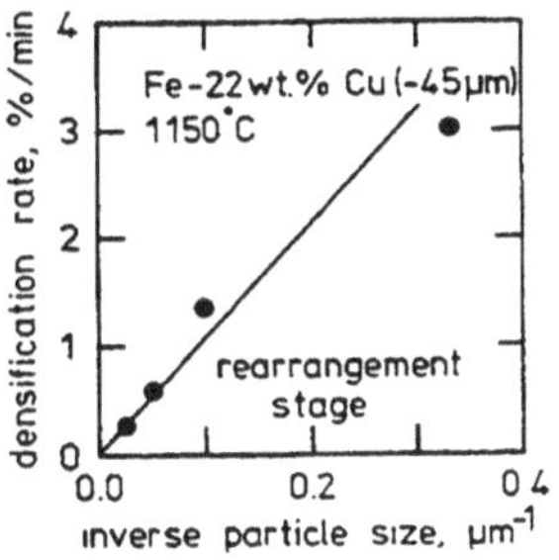

Figure 4.18 The densification rate in the rearrangement stage versus the particle size for Fe-Cu compacts, indicating an inverse particle size dependence (54).

sintering, concerns over dimensional control and uniform densification lead to high green densities. In cases where swelling occurs, a high green density contributes to greater swelling. Figure 4.20 illustrates the compaction pressure effect on densification in a Fe-Cu alloy. As the compaction pressure increases, the initial porosity is reduced from 40% at 0 MPa pressure to 7% at 785 MPa pressure, and the swelling increases. Although there is greater swelling at high compaction pressures, the final density is highest for 785 MPa.

There are combinations of mixed powders which give swelling on liquid formation. In such systems, the more uneven the initial distribution of the liquid, the greater the amount of swelling. Alternatively, homogeneous powder mixtures give better densification. Figure 4.21 shows this effect for W-Cu where the volume shrinkage is plotted versus the degree of mixing

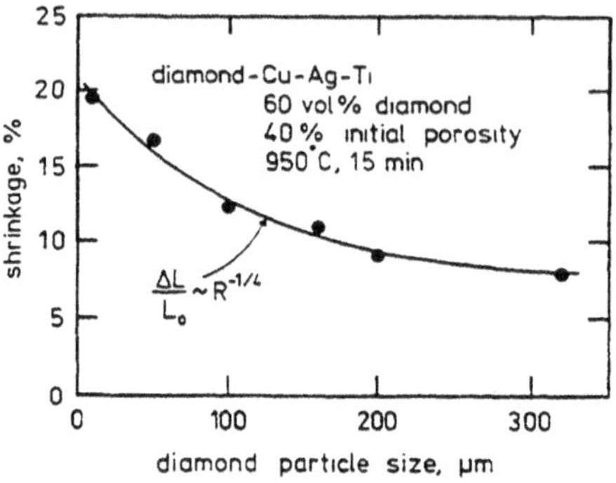

Figure 4.19 Rearrangement stage shrinkage for diamond powders of various sizes processed by liquid phase sintering in a Cu-Ag-Ti liquid (58).

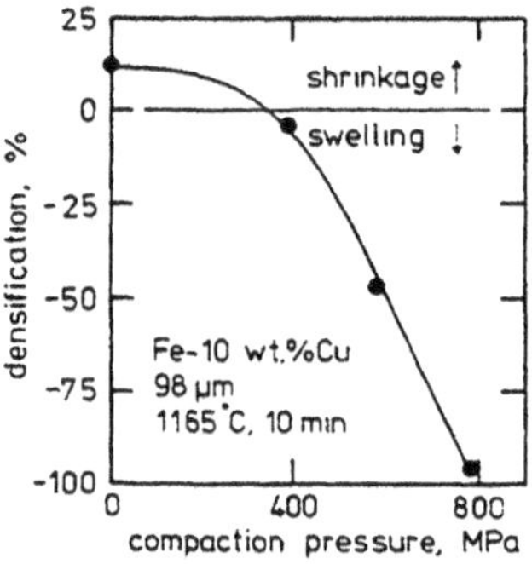

Figure 4.20 The swelling and shrinkage of Fe-Cu alloys as dependent on the compaction pressure under constant sintering conditions.

(37,60). The degree of mixing was measured by statistical sampling and ranged from mixed powders of equal size to coated tungsten powder. Coated powder gives the most homogeneous liquid and best densification. Besides differences in the degree of densification, there are also microstructural differences. Clustering as a secondary rearrangement process decreases as melt formation becomes more homogeneous. Uniform pore filling by the liquid requires a homogeneous distribution of liquid and uniform porosity. The initial pore size after melt formation decreases as the mixing and packing uniformity increase (53). For these reasons it is common to mill the ingredients in systems like WC-Co. Milling gives both a finer particle size and more homogeneous liquid. The best situation is obtained using coated powders where the liquid forms on all of the particles and at all of the contacts. Likewise, densification is more isotropic as the packing homogeneity

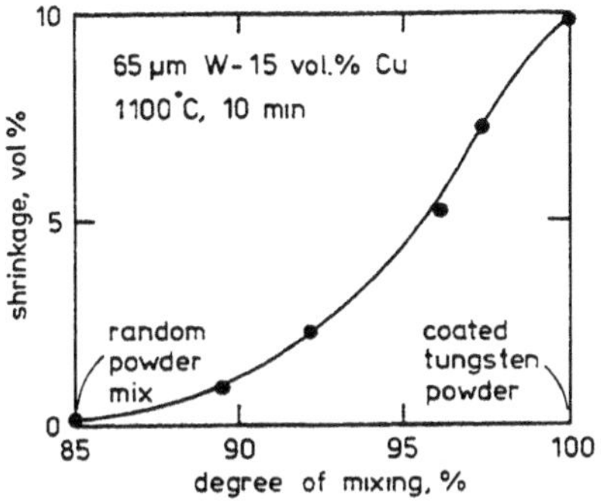

Figure 4.21 The volumetric shrinkage for W-Cu powder compacts with various degrees of mixing ranging from mixed powder to coated powder (60).

increases. Isotropic shrinkage is most important in the production of net shape components by powder processing techniques.

G. Pore Characteristics

There are two pore changes associated with the initial stage of liquid phase sintering. The first involves pore formation at prior additive particle sites in the microstructure (61). The second involves pore coarsening induced by particle rearrangement. In agreement with clustering observations, the mean pore size increases even as the total porosity decreases during the initial stage (59). Some two dimensional model studies have demonstrated pore coarsening due to unbalanced torques found in a random powder array (62). However, such events are less active in high density powder compacts. The capillary attraction for a liquid by the surrounding pore structure generates a pore at the particle sites where the liquid forms. The resulting pore may be difficult to remove during subsequent sintering, especially if the additive particle size is large. With prolonged sintering, pore refilling occurs if the dihedral angle and contact angle remain small (63). Because the initial pore size is roughly the same as the additive particle size, it is desirable to have small additive particle sizes. This is especially true with high green densities.

The size and distribution of pores in the microstructure at the end of the initial stage depend on the liquid distribution. Accordingly, factors like the powder packing and melt uniformity in the microstructure are influential. A low liquid solubility for the solid further creates an opportunity for pore formation and coarsening. Finally, the sintering atmosphere can affect pore evolution. Generally, vacuum provides the least difficulty. Gas can become trapped in the pores as liquid flows and can stabilize pores. However, a sintering atmosphere is often necessary to keep a chemical balance or to achieve oxide reduction.

H. Phase Diagram Concepts

The features associated with initial stage liquid phase sintering are not the equilibrium conditions. However, the equilibrium thermodynamic conditions provide a gauge of the driving forces effecting the system behavior. One key is the solubility ratio as illustrated in Figure 4.22. This figure shows the solubility ratio at the eutectic temperature for two extremes. Other factors affecting initial stage behavior are the amount and composition of liquid at the sintering temperature, formation of intermediate phases, and melting behavior. Several treatments have pointed out the importance of phase diagrams in understanding liquid phase sintering (15,56,64-66). An ideal binary phase diagram for liquid phase sintering is shown in Figure 4.22b. The important solubility and melting characteristics are evident. A typical sintering temperature would be slightly above the eutectic temperature, and a typical composition would be in the $L+\beta$ region at the sintering temperature. Various other simple diagrams are shown in Figure 4.23 along with the anticipated dimensional change on heating to the sintering temperature. In these cases a slow heating rate is assumed. For comparison, binary diagrams for various iron systems are shown in Figure 4.24. Swelling occurs with initial melt formation in systems like Fe-Cu and Fe-Sn (67). This is expected based on the phase diagram features and the concepts developed in this chapter. Densification during liquid phase sintering is associated with systems with a deep eutectic such as seen in the Fe-P and Fe-B systems. A steep liquidus for the base component is also favorable. Such a condition indicates more liquid for a given amount of additive, a high solubility of the base in the liquid, and substantial reduction in the sintering temperature. As an example, contrast the Fe-Cu and Fe-B binary diagrams. Boron as an additive is essentially insoluble in iron and has a deep eutectic with a steep liquidus. For the Fe-Cu system, the liquidus is very flat, there is a high solubility of Cu in Fe, and the concentration of Fe in the liquid is low. These factors are less favorable for liquid phase sintering. Comparative studies between these two systems show boron to be more effective in inducing densification of iron.

As developed in this chapter, several events occur during heating of mixed powders. Swelling is associated with additive dissolution in the base; both by solid state diffusion and after liquid formation. Shrinkage occurs when the base has a high solubility in the additive. On formation of the melt, the solubility of the base in the melt affects the dimensional change. A high base solubility in the melt induces densification by rearrangement. The appearance of an intermediate phase between the constituents leads to initial swelling with subsequent densification when the liquid forms. In general, simple systems, lacking intermediate compounds, with high solubility ratios and deep eutectics are most favorable for liquid phase sintering.

I. Contact Formation

Associated with rearrangement, reaction, swelling, and shrinkage in the initial stage of liquid phase sintering is the formation of new contacts. Contacts between the solid grains result in an increased contiguity and connectivity. Practical interest in liquid phase sintering is found at volume fractions of solid over approximately 50 percent where there is substantial interparticle contact (68). Without these contacts, the compact has no rigidity; thus, slumping, liquid phase run-out, and density gradients are common problems. As the liquid spreads, capillary forces create new contacts

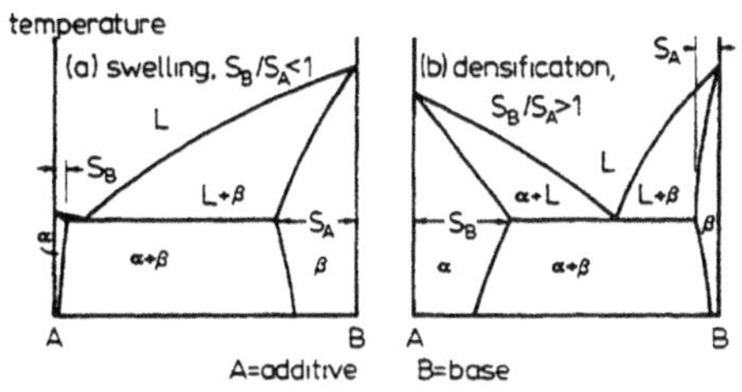

Figure 4.22 Model phase diagrams showing the solubility features at the eutectic temperature associated with swelling and densification.

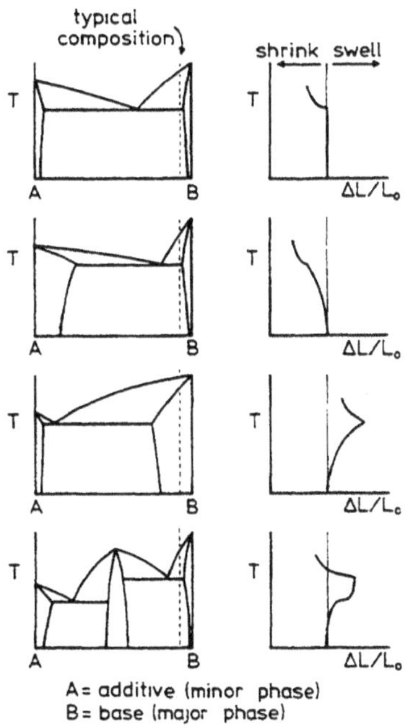

Figure 4.23 The effect of phase diagram characteristics on the dimensional change during heating for mixed powder systems.

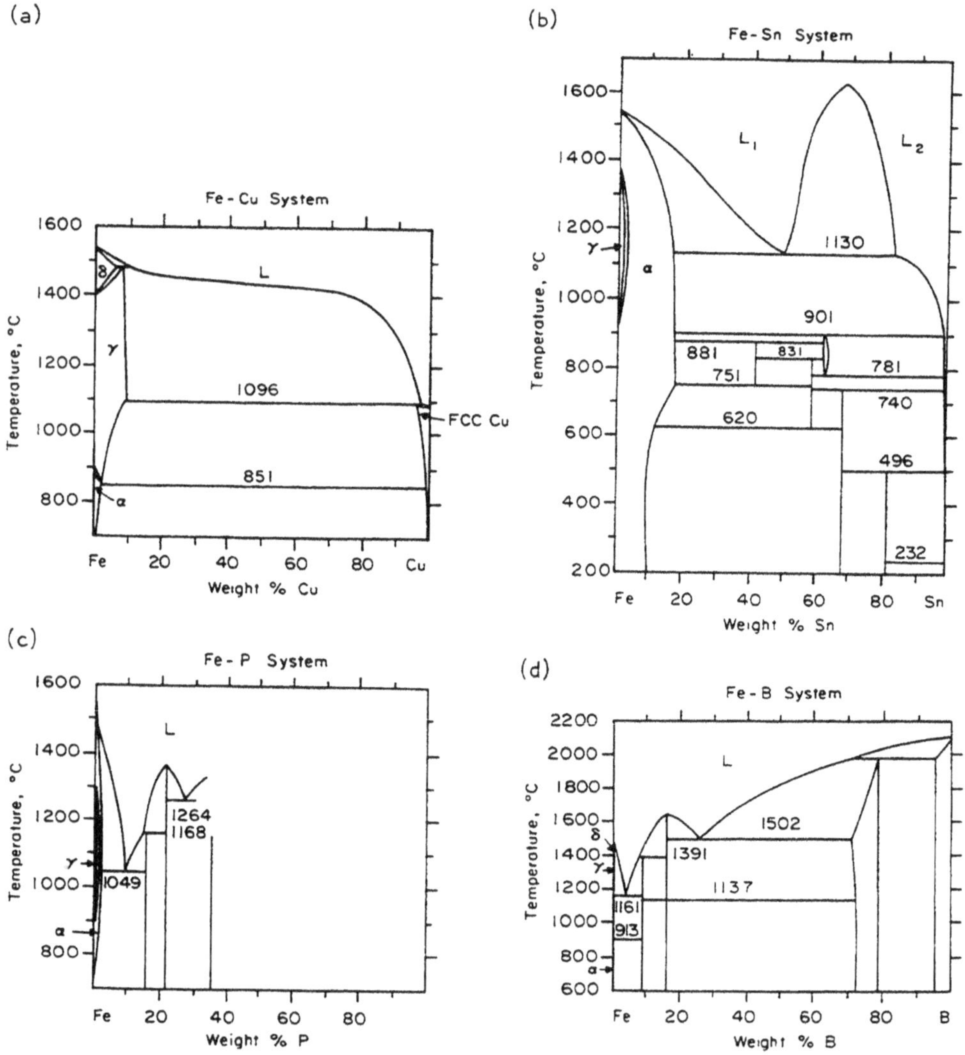

Figure 4.24 Binary phase diagrams for four iron-based liquid phase sintering systems; (a) Fe-Cu, (b) Fe-Sn, (c) Fe-P, and (d) Fe-B.

between particles which bond together to increase the compact rigidity. At this point the initial stage processes end and intermediate stage liquid phase sintering begins.

In high liquid content systems, contact formation is delayed, but occurs by settling of either the liquid or solid. The rate of settling increases as the solid and liquid densities differ. Besides settling, Brownian motion, and thermal agitation give contact formation (69). Contact formation is the precursor to grain coalescence and grain shape accommodation. Courtney (69,70) has treated contact formation for amorphous solids. His treatment shows the conditions where gravity or Brownian motion will provide appreciable contributions to contact formation. The stability of a random contact between two crystalline grains depends on the orientation of the interface. There is a small probability of a low angle grain boundary. Niemi et al. (71) estimate the probability of a random contact forming a low angle grain boundary as 0.005. For such contacts, rapid coalescence is anticipated.

At high volume fractions of solid or with high green densities, little motion of the grains in the liquid is possible. In these cases coalescence should not be important to liquid phase sintering beyond the first few minutes. Thus, in typical cases, contact formation by Brownian motion is not probable, and gravity induced contact is more probable.

J. Studies on Common Systems

In this section, the factors introduced in this chapter are used to explain the initial stage liquid phase sintering of three systems; Cu-Al, Fe-Cu, and W-Ni.

Experiments with Cu-Al show swelling during heating (20,32,72-74), as would be predicted from the binary phase diagram shown in Figure 4.25. With mixed powders the system undergoes an exothermic reaction, giving pore formation during heating. The final porosity increases with the initial porosity and amount of aluminum as shown in Figure 4.26. The aluminum additive diffuses into the surrounding copper prior to the formation of the first melt. Consequently, high final sintering temperatures are required to offset the swelling during heating. Studies on quenched microstructures show a reaction zone around the prior Al particle sites. The reaction gives swelling which increases in magnitude with the initial porosity and amount of Al. The model represented by Equation (4.7) shows good agreement with the experimental studies. The finer the aluminum particle size, the greater and faster the swelling. Because of the poor sintering response, Cu-Al is not a useful system. It is interesting that Al with Cu additions exhibits densification behavior, most likely due to a favorable solubility ratio.

A system of practical interest is Fe-Cu, especially with carbon additions. The initial stage sintering behavior has been studied on several occasions (1,4,25,44,75-81). The binary phase diagram for the Fe-Cu system is shown in Figure 4.24a. In this case there is initial melt induced swelling as shown in Figure 4.27a. Pores form at the prior Cu particle sites. A secondary stage of swelling occurs if the iron particle are polycrystalline, due to grain boundary penetration as illustrated in Figure 4.27b. The amount of swelling is dependent on the green density, amount of copper, particle sizes, internal powder porosity, copper distribution, and amount of carbon.

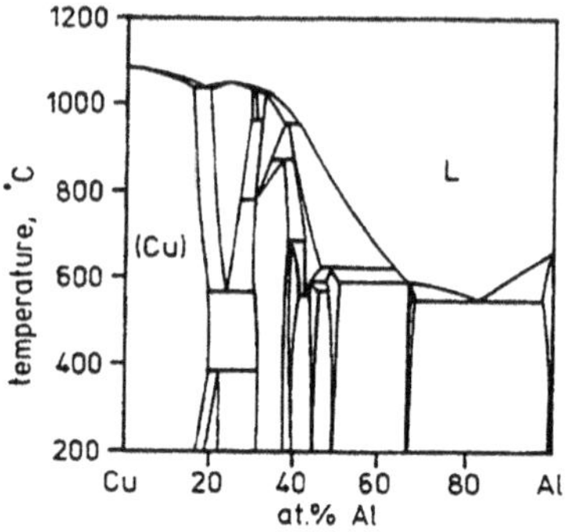

Figure 4.25 The copper-aluminum binary phase diagram, a system which undergoes swelling during heating to a liquid phase sintering temperature.

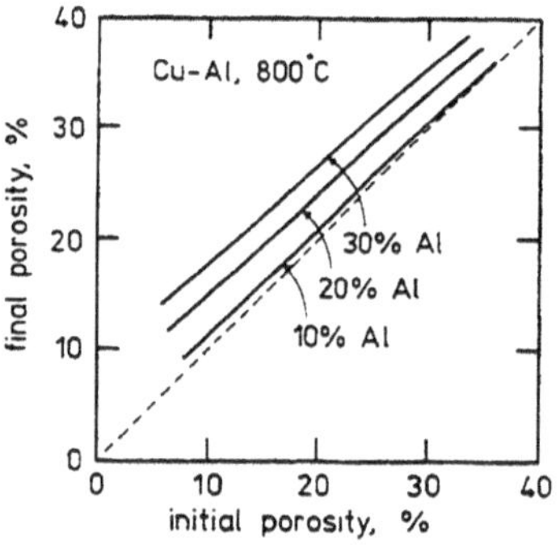

Figure 4.26 The final porosity after sintering at 800°C for Cu-Al powder mixtures versus the Al content and initial porosity (72).

For Fe-Cu alloys, the green density effect is very large; a high compaction pressure contributes to swelling as illustrated in Figure 4.28. Loose powder will densify with liquid formation. Alternatively, a high compaction pressure gives swelling and less rearrangement densification. An increase in the green density produces greater interparticle contact. During the initial stage, the melt penetrates these contacts to establish an equilibrium dihedral angle. Microstructural analysis shows the maximum in swelling corresponds to the point where the dihedral angle is at a minimum (75). Furthermore, as the Fe particle size increases, swelling goes through a peak. At small particle sizes the capillary forces give shrinkage. At large Fe particle sizes, there are fewer interparticle regions for Cu penetration, thus less swelling is observed.

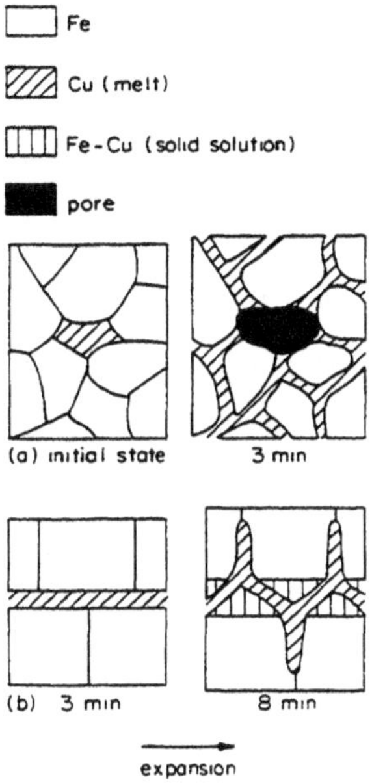

Figure 4.27 Two mechanisms of swelling in the Fe-Cu system (81). Swelling occurs because of particle separation (a) and grain boundary penetration (b).

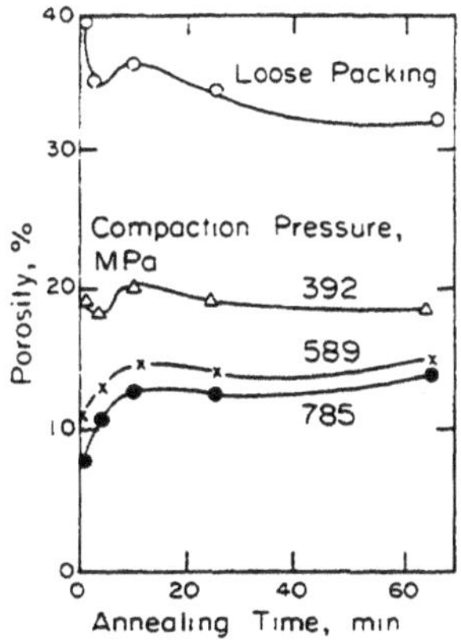

Figure 4.28 The sintered porosity versus time at 1165°C for Fe-10 wt.% Cu compacts with varying compaction states (81).

Carbon additions are useful for controlling dimensions during the initial stage as shown in Figure 4.29. Carbon has a high solubility in Fe at temperatures over 910°C. The diffusion of carbon into iron increases the dihedral angle as shown in Figure 4.30, and thereby inhibits swelling due to grain boundary penetration by molten copper at high green densities. An increase in the internal porosity of the Fe powder or a decrease in the initial green density reduces the swelling. This results from the capillary attraction of the melt to the fine pores which inhibits melt penetration between solid grains. The overall process of initial stage densification is very dependent on the Fe particle size. Fine Fe particle sizes give the highest densification.

The W-Ni system provides a high solubility ratio and rapid densification prior to melt formation. This system is an ideal candidate for liquid phase sintering because of the high sintering temperature for pure tungsten and the dramatic densification possible from the addition of small quantities of nickel (10,37,38,43,66,82). Figure 4.31 gives the W-Ni phase diagram, showing the desired attributes of liquid formation and a high solubility ratio. The favorable solubility ratio gives considerable solid state densification for W-Ni powder mixtures. Indeed, near full density is possible without forming a liquid. Figure 4.32 shows sintered density as a percentage of theoretical density versus sintering temperature for W-Ni alloys with two different tungsten particle sizes. Substantial densification occurs prior to the formation of the first liquid. This densification greatly exceeds that attained by tungsten without the nickel additive. Only for the coarser powder is there continued densification at temperatures involving liquid phase sintering. On formation of the liquid, any polycrystalline tungsten particles are fragmented. Fragmentation combines with solution-reprecipitation to give rapid rearrangement and grain shape accommodation (64). Examination of the pore structure shows the mean pore size initially increases after liquid formation even though the porosity decreases (48). As the particle size of the tungsten increases, there is less solid state sintering and less rearrangement on melt formation.

K. Summary of Initial Stage Events

The initial stage of liquid phase sintering typically ends within the first 10 minutes after melt formation. Shrinkage will occur during heating if diffusion and solubility favor base flow into the additive. Swelling will occur if there is a small solubility ratio. The rate of densification and the degree of densification both increase with solid solubility in the liquid. Thus, the mere presence of a liquid is insufficient for densification.

After the liquid forms, the concern is with rearrangement. The solid particles repack under capillary forces from a wetting liquid. There are other processes important to the later stages of liquid phase sintering which are active during the initial stage. However, these processes are slow and less significant in the first few minutes after melt formation.

Rearrangement occurs in two stages. Primary rearrangement involves the original particles. The spreading of a wetting liquid between particles causes capillary induced contact formation. Primary rearrangement is rapid and depends on the melt spreading rate. Secondary rearrangement involves clustering and penetration. Clustering results from the expansion of the melt zone through the unwetted particles. Penetration leads to disintegration of the solid and subsequent rearrangement of fragments.

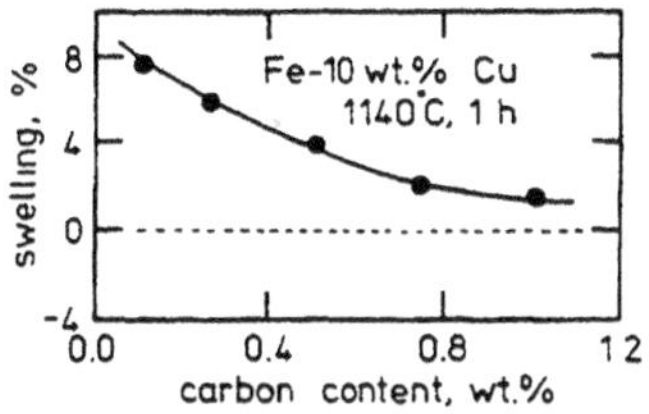

Figure 4.29 Swelling of Fe-Cu compacts is decreased by carbon additions because of retarded grain boundary penetration (44).

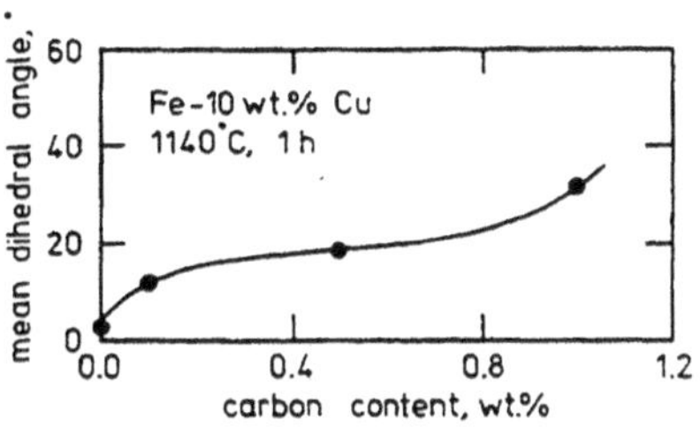

Figure 4.30 The mean dihedral angle of Fe-Cu compacts versus the carbon content. Swelling decreases as the carbon content is increased (44).

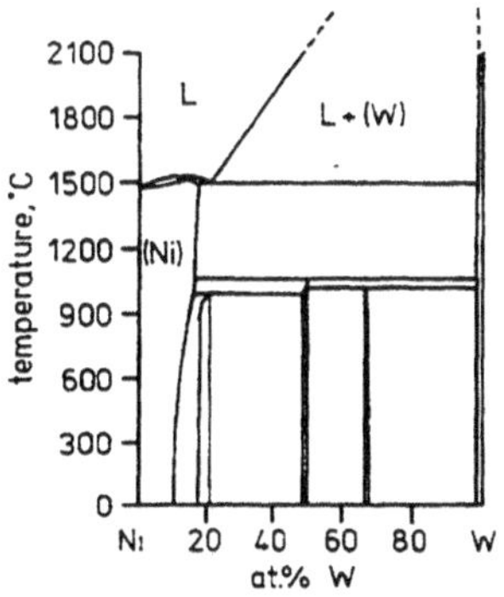

Figure 4.31 The binary phase diagram for Ni-W, a system which exhibits considerable densification during liquid phase sintering.

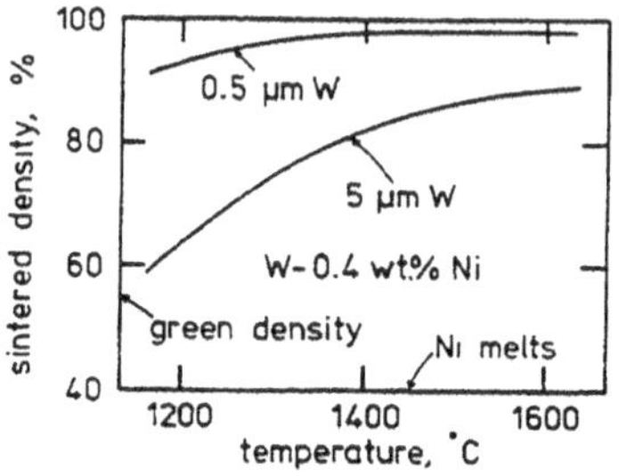

Figure 4.32 Sintered density versus sintering temperature for W-Ni compacts with two different particle sizes of W (82). Densification occurs prior to formation of the first liquid.

The most favorable features for rearrangement are a congruent melting liquid, wetting, homogeneous liquid distribution, single crystal solid grains, small particle sizes, solid solubility in the liquid, and a large solubility ratio. A high volume fraction of liquid decreases the viscosity of the solid-liquid-pore ensemble but also decreases the capillary force. Thus, an intermediate quantity of liquid is most practical where viscosity and capillarity effects are optimized. A high green density inhibits rearrangement and often causes swelling. A high sintering temperature is generally useful for initial stage processes since it increases the solid solubility in the liquid and decreases the liquid viscosity.

Pore creation at prior additive particle sites is common during heating to the initial melt formation temperature. The pores result from reactivity, solubility, and diffusivity processes which lead to imbalanced mass flow from the additive to the surrounding matrix. The swelling events can be minimized by homogeneous mixing of the additive. It is clear that solubility is a major concern for initial stage liquid phase sintering. Phase diagrams offer the first basis for examining potential systems.

L. References

1. H. S. Cannon and F. V. Lenel, "Some Observations on the Mechanism of Liquid Phase Sintering," *Plansee Proceedings*, F. Benesovsky (ed.), Metallwerk Plansee, Reutte, Austria, 1953, pp.106-121.
2. W. D. Kingery, "Densification During Sintering in the Presence of a Liquid Phase. 1. Theory," *J. Appl. Phys.*, 1959, vol.30, pp.301-306.
3. V. N. Eremenko, Y. V. Naidich, and I. A. Lavrinenko, *Liquid Phase Sintering*, Consultants Bureau, New York, NY, 1970.
4. B. E. Magee and J. Lund, "Mechanisms of Liquid-Phase Sintering in Iron-Copper Powder Compacts," *Z. Metallkde.*, 1976, vol.67, pp.596-602.
5. F. V. Lenel, "Sintering in the Presence of a Liquid Phase," *Trans. AIME*, 1948, vol.175, pp.878-896.
6. T. J. Whalen and M. Humenik, "Sintering in the Presence of a Liquid Phase," *Sintering and Related Phenomena*, G. C. Kuczynski, N. Hooton, and C. Gibbon (eds.), Gordon and Breach, 1967, New York, NY, pp.715-74.

7. B. Meredith and D. R. Milner, "The Liquid Phase Sintering of Titanium Carbide," *Powder Met.*, 1976, vol.19, pp.162-170.
8. V. V. Panichkina, M. M. Sirotyuk, and V. V. Skorokhod, "Liquid-Phase Sintering of Very Fine Tungsten-Copper Powder Mixtures," *Soviet Powder Met. Metal Ceram.*, 1982, vol.21, pp.447-450.
9. R. F. Snowball and D. R. Milner, "Densification Processes in the Tungsten Carbide-Cobalt System," *Powder Met.*, 1968, vol.11, pp.23-40.
10. W. J. Huppmann, H. Riegger, and G. Petzow, "Liquid Phase Sintering of the Model System W-Ni," *Sintering - New Developments*, M. M. Ristic (ed.), Elsevier Scientific, Amsterdam, Netherlands, 1979, pp.272-278.
11. W. J. Huppmann and G. Petzow, "Particle Rearrangement During Liquid Phase Sintering of Several Carbide-Metal Combinations," *Modern Developments in Powder Metallurgy*, vol.9, H. H. Hausner and P. W. Taubenblat (eds.), Metal Powder Industries Federation, Princeton, NJ, 1977, pp.77-89.
12. F. Aldinger, "Controlled Porosity by an Extreme Kirkendall Effect," *Acta Met.*, 1974, vol.22, pp.923-928.
13. L. Froschauer and R. M. Fulrath "Direct Observation of Liquid-Phase Sintering in the System Tungsten Carbide-Cobalt," *J. Mater. Sci.*, 1976, vol.11, pp.142-149.
14. A. P. Savitskii, E. S. Kim, and L. S. Martsunova, "Compact Shrinkage During Liquid-Phase Sintering," *Soviet Powder Met. Metal Ceram.*, 1980, vol.19, pp.593-596.
15. A. P. Savitskii, L. S. Martsunova, and M. A. Emelyanova, "Compact Property Changes in Liquid-Phase Sintering due to Diffusional Interaction Between Phases," *Soviet Powder Met. Metal Ceram.*, 1981, vol.20, pp.4-9.
16. A. P. Savitskii and L. S. Martsunova, "Effect of Solid-State Solubility on the Volume Changes Experienced by Aluminum During Liquid-Phase Sintering," *Soviet Powder Met. Metal Ceram.*, 1977, vol.16, pp.333-337.
17. A. P. Savitskii, N. N. Burtsev, and L. S. Martsunova, "Volume Changes Experienced by Al-Zn Compacts During Liquid-Phase Sintering," *Soviet Powder Met. Metal Ceram.*, 1982, vol.21, pp.760-764.
18. A. P. Savitskii, "Some Characteristic Features of the Sintering of Binary Systems," *Soviet Powder Met. Metal Ceram.*, 1980, vol.19, pp.488-493.
19. A. P. Savitskii and N. N. Burtsev, "Compact Growth in Liquid Phase Sintering," *Soviet Powder Met. Metal Ceram.*, 1979, vol.18, pp.96-102.
20. K. V. Savitskii, V. I. Itin, Y. I. Kozlov, and A. P. Savitskii, "The Effect of the Particle Size of Aluminum Powder on the Sintering of a Cu-Al Alloy in the Presence of a Liquid Phase," *Soviet Powder Met. Metal Ceram.*, 1965, vol.4, pp.886-890.
21. E. Peissker, "Pressing and Sintering Characteristics of Powder Mixtures for Sintered Bronze 90/10 Containing Different Amounts of Free Tin," *Modern Developments in Powder Metallurgy*, vol.7, H. H. Hausner and W. E. Smith (eds.), Metal Powder Industries Federation, Princeton, NJ, 1974, pp.597-614.
22. B. Rieger, W. Schatt, and C. Sauer, "Combined Mechanical Activation and Sintering with a Short-Time Occurance of a Liquid Phase," *Inter. J. Powder Met. Powder Tech.*, 1983, vol.19, pp.29-41.
23. D. J. Lee and R. M. German, "Sintering Behavior of Iron-Aluminum Powder Mixes," *Inter. J. Powder Met. Powder Tech.*, 1985, vol.21, pp.9-21.
24. K. S. Hwang and R. M. German, "High Density Ferrous Components by Activated Sintering," *Processing of Metal and Ceramic Powders*, R. M.

German and K. W. Lay (eds.), The Metallurgical Society, Warrendale, PA, 1982, pp.295-310.
25. V. B. Phadke and B. L. Davies, "Comparing Diffusion and Penetration Theories of Growth in P/M Iron-Copper Alloys," *Inter. J. Powder Met. Powder Tech.*, 1977, vol.13, pp.253-258.
26. F. J. Esper, K. H. Friese, and R. Zeller, "Sintering Reaction and Radial Compressive Strength of Iron-Tin and Iron-Copper-Tin Powder Compacts," *Inter. J. Powder Met.*, 1969, vol.5, no.3, pp.19-32.
27. B. K. Kiebach, W. Schatt, and G. Jangg, "Titanium-Alloyed Sintered Steels," *Powder Met. Inter.*, 1984, vol.16, pp.207-212.
28. M. Lejbrandt and W. Rutkowski, "The Effect of Grain Size of Nickel Activating the Sintering of Molybdenum," *Planseeber. Pulvermetall.*, 1977, vol.25, pp.3-12.
29. A. P. Savitskii and N. N. Burtsev, "Effect of Powder Particle Size on the Growth of Titanium Compacts During Liquid-Phase Sintering with Aluminum," *Soviet Powder Met. Metal Ceram.*, 1981, vol.20, pp.618-621.
30. I. Amato, "On the Mechanism of Activated Sintering of Tungsten Powders," *Mater. Sci. Eng.*, 1972, vol.10, pp.15-22.
31. G. C. Kuczynski, L. Abernethy, and J. Allen, "Sintering Mechanisms of Aluminum Oxide," *Kinetics of High Temperature Processes*, W. D. Kingery (ed.), John Wiley, New York, NY, 1959, pp.163-172.
32. K. V. Savitskii, V. I. Itin, and Y. I. Kozlov, "The Mechanism of Sintering of Copper-Aluminum Powder Alloys in the Presence of a Liquid Phase," *Soviet Powder Met. Metal Ceram.*, 1966, vol.5, pp.4-9.
33. R. J. Nelson and D. R. Milner, "Liquid-Flow Densification in the Tungsten Carbide-Copper System," *Powder Met.*, 1971, vol.14, pp.39-63.
34. S. Pejovnik, D. Kolar, W. J. Huppmann, and G. Petzow, "Sintering of Alumina in Presence of Liquid Phase," *Sintering - New Developments*, M. M. Ristic (ed.), Elsevier Scientific, Amsterdam, Netherlands, 1979, pp.285-292.
35. H. Riegger, J. A. Pask, and H. E. Exner, "Direct Observation of Densification and Grain Growth in a W-Ni Alloy," *Sintering Processes*, G. C. Kuczynski (ed.), Plenum Press, New York, NY, 1980, pp.219-233.
36. T. Kosmac, D. Kolar, M. Komac, M. Trontelj, and M. Brloznik, "The Influence of VC Additions on the Sintering of Ta-Carbonitride Based Hard Metals," *Sci. Sintering*, 1979, vol.11 (special supplement), pp.97-104.
37. W. J. Huppmann, "Sintering in the Presence of a Liquid Phase," *Sintering and Catalysis*, G. C. Kuczynski (ed.), Plenum Press, New York, NY, 1975, pp.359-378.
38. W. J. Huppmann and G. Petzow, "The Elementary Mechanisms of Liquid Phase Sintering," *Sintering Processes*, G. C. Kuczynski (ed.), Plenum Press, New York, NY, 1980, pp.189-201
39. I. A. Aksay, C. E. Hoge, and J. A. Pask, "Phase Distribution in Solid-Liquid-Vapor Systems," *Surfaces and Interfaces of Glass and Ceramics*, V. D. Frechette, W. C. Lacourse, and V. L. Burdick (eds.), Plenum Press, New York, NY, 1974, pp.299-321.
40. W. Beere, " A Unifying Theory of the Stability of Penetrating Liquid Phases and Sintering Pores," *Acta Met.*, 1975, vol.23, pp.131-138.
41. S. R. Jurewicz and E. B. Watson, "Distribution of Partial Melt in a Felsic System: The Importance of Surface Energy," *Contrib. Mineral. Pwtrol.*, 1984, vol.85, pp.25-29.
42. T. Schmitt, M. Schreiner, E. Lassner, and B. Lux, "Uber das Diskontinuierliche Kornwachstum in WC/Co Hartmetallen," *Z. Metallkde.*,

1983, vol.74, pp.496-499.
43. W. J. Huppmann and H. Riegger, "Liquid Phase Sintering of the Model System W-Ni," *Inter. J. Powder Met. Powder Tech.*, 1977, vol.13, pp.243-247.
44. S. J. Jamil and G. A. Chadwick, "Investigation and Analysis of Liquid Phase Sintering of Fe-Cu and Fe-Cu-C Compacts," *Proceedings Sintering Theory and Practice Conference,* The Metals Society, London, UK, 1984, pp.13.1-13.14.
45. K.-S. Hwang, "Analysis of the Initial Stage of Sintering in the Solid and Liquid Phase," Ph.D. Thesis, Rensselaer Polytechnic Institute, Troy, NY, 1984.
46. R. B. Heady and J. W. Cahn, "An Analysis of the Capillary Forces in Liquid-Phase Sintering of Spherical Particles," *Metall. Trans.*, 1970, vol.1, pp.185-189.
47. Y. V. Naidich, I. A. Lavrinenko, and V. Y. Petrishchev, "Study on the Capillary Adhesive Forces Between Solid Particles with a Liquid Layer at the Points of Contact. 1. Spherical Particles," *Soviet Powder Met. Metal Ceram.*, 1965, vol.4, pp.129-133.
48. W. J. Huppmann and R. Riegger, "Modelling of Rearrangement Processes in Liquid Phase Sintering," *Acta Met.*, 1975, vol.23, pp.965-971.
49. V. Smolej and S. Pejovnik, "Some Remarks on the Driving Force for Liquid-Phase Sintering," *Z. Metallkde.*, 1976, vol.67, pp.603-605.
50. G. Petzow and W. J. Huppmann, "Flussigphasensintern Verdichtung und Gefugeausbildung," *Z. Metallkde.*, 1976, vol.67, pp.579-590.
51. Z. Panek, "The Optimization of Liquid Amount During Sintering," *Sci. Sintering*, 1984, vol.16, pp.13-20.
52. J. W. Cahn and R. B. Heady, "Analysis of Capillary Forces in Liquid-Phase Sintering of Jagged Particles," *J. Amer. Ceramic Soc.*, 1970, vol.53, pp.406-409.
53. B. Meredith and D. R. Milner, "Densification Mechanisms in the Tungsten Carbide-Cobalt System," *Powder Met.*, 1976, vol.19, pp.38-45.
54. W. D. Kingery and M. D. Narasimhan, "Densification During Sintering in the Presence of a Liquid Phase. 2. Experimental," *J. Appl. Phys.*, 1959, vol.30, pp.307-310.
55. W. D. Kingery, E. Niki, and M. D. Narasimhan, "Sintering of Oxide and Carbide-Metal Compositions in Presence of a Liquid Phase," *J. Amer. Ceramic Soc.*, 1961, vol.44, pp.29-35.
56. D. L. Johnson and I. B. Cutler, "The Use of Phase Diagrams in the Sintering of Ceramics and Metals," *Phase Diagrams*, vol.2, A. M. Alper (ed.), Academic Press, New York, NY, 1970, pp.265-291.
57. M. A. Fortes, "The Kinetics of Powder Densification due to Capillary Forces," *Powder Met. Inter.*, 1982. vol.14, pp.96-100.
58. Y. V. Naidich, I. A. Lavrinenko, and V. A. Evdokimov, "Densification During Liquid-Phase Sintering in Diamond-Metal Systems," *Soviet Powder Met. Metal Ceram.*, 1972, vol.11, pp.715-718.
59. W. J. Huppmann, H. Riegger, W. A. Kaysser, V. Smolej, and S. Pejovnik, "The Elementary Mechanisms of Liquid Phase Sintering I. Rearrangement," *Z. Metallkde.*, 1979, vol.70, pp.707-713.
60. W. J. Huppmann and W. Bauer, "Characterization of the Degree of Mixing in Liquid-Phase Sintering Experiments," *Powder Met.*, 1975, vol.18, pp.249-258.
61. O. J. Kwon and D. N. Yoon, "The Liquid Phase Sintering of W - Ni," *Sintering Processes*, G. C. Kuczynski (ed.), Plenum Press, New York, NY, 1980, pp.203-218.

62. V. Smolej, S. Pejovnik, and W. A. Kaysser, "Rearrangement During Liquid Phase Sintering of Large Particles," *Powder Met. Inter.*, 1982, vol.14, pp.34-36.
63. O. J. Kwon and D. N. Yoon, "Closure of Isolated Pores in Liquid Phase Sintering of W-Ni," *Inter. J. Powder Met. Powder Tech.*, 1981, vol.17, pp.127-133.
64. W. J. Huppmann, W. A. Kaysser, D. N. Yoon, and G. Petzow, "Progress in Liquid Phase Sintering," *Powder Met. Inter.*, 1979, vol.11, pp.50-51.
65. P. E. Zovas, R. M. German, K. S. Hwang, and C. J. Li, "Activated and Liquid-Phase Sintering - Progress and Problems," *J. Metals*, 1983, vol.35, no.1, pp.28-33.
66. C. J. Li and R. M. German, "Enhanced Sintering of Tungsten - Phase Equilibria Effects on Properties," *Inter. J. Powder Met. Powder Tech.*, 1984, vol.20, pp.149-162.
67. R. M. German and K. A. D'Angelo, "Enhanced Sintering Treatments for Ferrous Powders," *Inter. Metals Rev.*, 1984, vol.29, pp.249-272.
68. A. N. Niemi and T. H. Courtney, "The Temperature Dependence of Grain Boundary Wetting in Liquid Phase Sintered Fe-Cu Alloys," *Metall. Trans. A*, 1981, vol.12A, pp.1987-1988.
69. A. N. Niemi and T. H. Courtney, "Settling in Solid-Liquid Systems with Specific Application to Liquid Phase Sintering," *Acta Met.*, 1983, vol.31, pp.1393-1401.
70. T. H. Courtney, "Densification and Structural Development in Liquid Phase Sintering," *Metall. Trans. A*, 1984, vol.15A, pp.1065-1074.
71. A. N. Niemi, L. E. Baxa, J. K. Lee, and T. H. Courtney, "Coalescence Phenomena in Liquid Phase Sintering - Conditions and Effects on Microstructure," *Modern Developments in Powder Metallurgy*, vol.12, H. H. Hausner, H. W. Antes, and G. D. Smith (eds.), Metal Powder Industries Federation, Princeton, NJ, 1981, pp.483-495.
72. A. P. Savitskii, M. A. Emelyanova, and N. N. Burtsev, "Volume Changes Exhibited by Cu-Al Compacts During Liquid-Phase Sintering," *Soviet Powder Met. Metal Ceram.*, 1982, vol.21, pp.373-378.
73. W. Kehl and H. F. Fischmeister, "Liquid Phase Sintering of Al-Cu Compacts," *Powder Met.*, 1980, vol.23, pp.113-119.
74. L. S. Martsunova, A. P. Savitskii, E. N. Ushakova, and B. I. Matveev, "Sintering of Aluminum with Copper Additions," *Soviet Powder Met. Metal Ceram.*, 1973, vol.12, pp.956-959.
75. D. Berner, H. E. Exner, and G. Petzow, "Swelling of Iron-Copper Mixtures During Sintering and Infiltration," *Modern Developments in Powder Metallurgy*, vol.6, H. H. Hausner and W. E. Smith (eds.), Metal Powder Industries Federation, Princeton, NJ, 1974, pp.237-250.
76. K. Tabeshfar and G. A. Chadwick, "Dimensional Changes During Liquid Phase Sintering of Fe-Cu Compacts," *Powder Met.*, 1984, vol.27, pp.19-24.
77. L. Froschauer and R. M. Fulrath, "Direct Observation of Liquid-Phase Sintering in the System Iron-Copper," *J. Mater. Sci.*, 1975, vol.10, pp.2146-2155.
78. W. A. Kaysser, S. Takajo, and G. Petzow, "Skeleton Dissolution and Skeleton Formation During Liquid Phase Sintering of Fe-Cu," *Modern Developments in Powder Metallurgy*, vol.12, H. H. Hausner, H. W. Antes, and G. D. Smith (eds.), Metal Powder Industries Federation, Princeton, NJ, 1981, pp.474-481.
79. K. Tabeshfar and G. A. Chadwick, "The Role of Powder Characteristics

and Compacting Pressure in Liquid Phase Sintering of Fe-Cu Compacts," *Proceedings P/M-82*, Associazione Italiana di Metallurgia, Milano, Italy, 1982, pp.693-700.
80. T. Krantz, "Effect of Density and Composition on the Dimensional Stability and Strength of Iron-Copper Alloys," *Inter. J. Powder Met.*, 1969, vol.5, no.3, pp.35-43.
81. W. A. Kaysser and G. Petzow, "Recent Conceptions on Liquid Phase Sintering," *Proceedings Sintering Theory and Practice Conference*, The Metals Society, London, UK, 1984, pp.10.1-10.6.
82. C. Li and R. M. German, "The Properties of Tungsten Processed by Chemically Activated Sintering," *Metall. Trans. A*, 1983, vol.14A, pp.2031-2041.

CHAPTER FIVE

Intermediate Stage Processes: Solution-Reprecipitation

A. Characteristic Features

During the period when the liquid forms and spreads, rearrangement events happen rapidly. Although solution and reprecipitation of the solid occurs concurrently with rearrangement, the rearrangement events dominate the early response. During the second stage of liquid phase sintering rearrangement ends and solution-reprecipitation processes become dominant. Solution-reprecipitation requires solid solubility in the liquid. It is characterized by grain growth, dissolution of small grains, grain rounding, densification, and development of a rigid skeleton of solid. The two main concerns are with elimination of residual porosity and the concomitant microstructural coarsening. Both processes are interrelated and depend on essentially the same kinetic steps. For this treatment the solution-reprecipitation controlled densification event will be treated independently from microstructural coarsening. The reader is cautioned that this separation of events is artificial and is intended to ease study of liquid phase sintering. Actually, microstructural coarsening occurs simultaneously with solution-reprecipitation controlled densification.

The rate of densification in the intermediate stage predominantly depends on the rate of mass transfer through the liquid. Table 5.1 lists the main events associated with the intermediate stage. As noted in this table, densification, contact flattening, neck growth, and microstructural coarsening (grain growth) are all evident in the intermediate stage. Figure 5.1 shows the changes in porosity, grain size, number of grains, neck size, number of necks, and mean free distance (grain separation) for Fe-20 wt.% Cu sintered at 1150°C (1). The porosity, number of grains, and number of necks decrease with time, while the other features increase. Densification during the intermediate stage depends on material flow through the liquid. The concurrent microstructural changes, such as grain coarsening and shape accommodation, lower the total system energy by the elimination of interfacial area. Thus, pores shrink while grains grow. Figure 5.2 shows an example of a microstructure typical to the intermediate stage. This is a sample of Fe-2% C sintered at 1165°C for one hour. Grain shape accommodation and residual porosity are evident in the microstructure. The amount and size of the pores decrease as the grain size increases. At the end of the intermediate stage, pores have either been eliminated or stabilized by a trapped internal atmosphere. Additionally, the grain structure has formed a rigid skeleton which retards further densification. The final stage of liquid phase sintering

TABLE 5.1

Intermediate Stage Densification And Coarsening

MECHANISMS	contact flattening	densification
	dissolution of fine grains	densification and coarsening
	coalescence	coarsening
CONTROLLING STEPS	liquid diffusion	through liquid
	solid diffusion	along grain contacts
	reaction	solid-liquid interface
NECK GROWTH	none	zero dihedral angle
	extensive	large dihedral angle

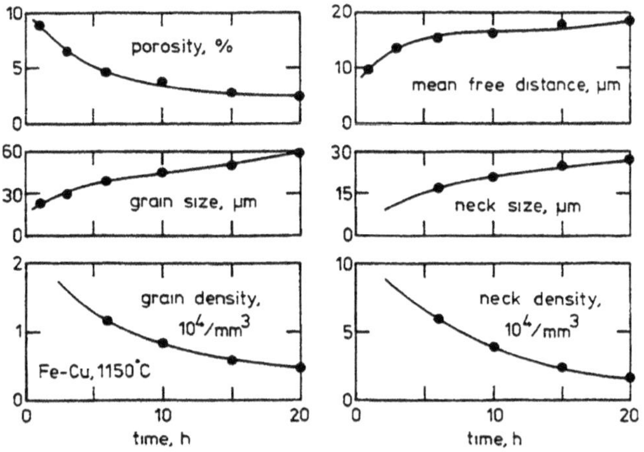

Figure 5.1 The several characteristic changes during intermediate stage liquid phase sintering as illustrated for Fe-20 wt.% Cu sintered at 1150°C (1).

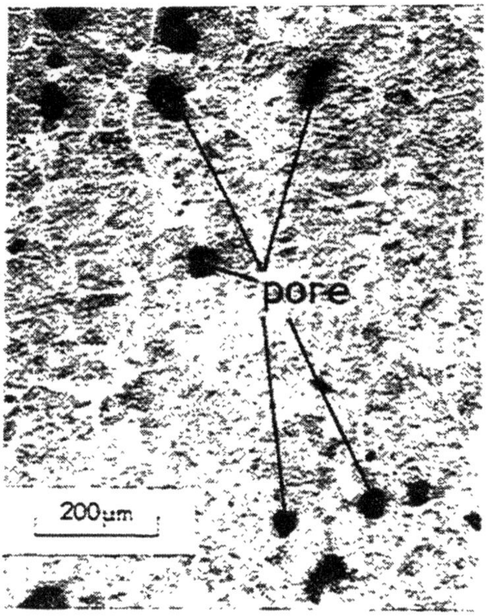

Figure 5.2 Microstructure typical to intermediate stage sintering; Fe-2 wt.% C sintered at 1165°C for 1 hour (photo courtesy of J. W. Dunlap).

corresponds to the period where grain growth continues without significant densification. Chapter 6 treats the simultaneous grain growth process in detail, while in this chapter the densification events associated with solution-reprecipitation are discussed.

B. Grain Shape Accommodation

In most useful liquid phase sintering systems, the quantity of liquid is insufficient to fill all pore space on melt formation. As a consequence, after rearrangement there is still porosity in the compact. Grain shape accommodation gives a higher solid packing density. As seen in Figure 5.3, the solid grains have deviated from a spherical (minimum energy) shape to fill space better. This more efficient solid space filling releases liquid to fill pores. For a given grain volume, this grain shape gives a higher solid-liquid surface area; however, the elimination of pores and the associated surface energy provides for a net energy decrease (2). The grains attain better packing by selective dissolution of the solid with reprecipitation at points in the microstructure removed from the grain contacts. During this solution-reprecipitation process, transport is through the high diffusivity liquid surrounding the solid grains. For solution-reprecipitation to be active, solid solubility in the liquid is necessary. Besides solution-reprecipitation, coalescence of small grains with contacting large grains also contributes to grain coarsening and shape accommodation (3). It is characteristic of the intermediate stage of

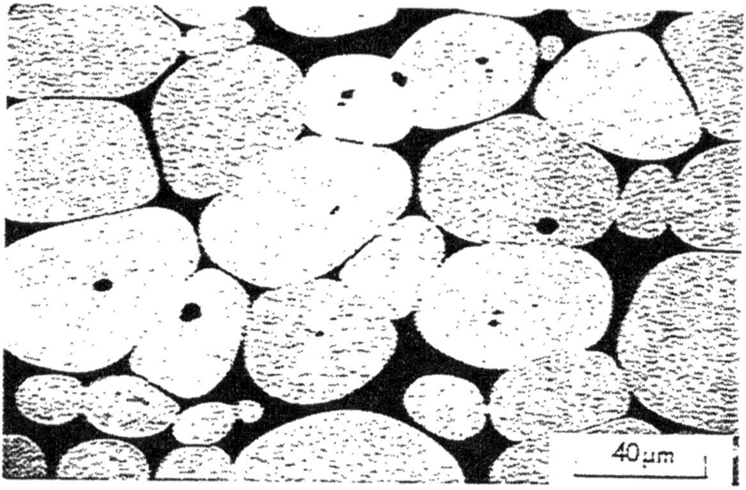

Figure 5.3 Grain shape accommodation of tungsten grains in a W-Ni-Fe heavy alloy after sintering at 1480°C for 1 hour.

liquid phase sintering that grain growth, shape accommodation, and densification occur simultaneously.

The net energy must decrease during solution-reprecipitation; thus, the reduction in surface energy due to pore filling has to exceed the surface energy increase due to grain shape distortion from a sphere. A compact with grain shape accommodation is not at the lowest energy condition. This has been demonstrated by Kaysser et al. (4). A full density compact with grain shape accommodation will take up liquid when it is immersed in a liquid of the equilibrium concentration. The added liquid allows the solid-liquid interface to relax to a spherical grain shape. Park et al. (5) define the force causing grain shape relaxation on immersion in an excess of liquid as the sphering force. Their calculations show the the sphering force is inversely dependent on the volume fraction of liquid. In turn, the degree of shape accommodation also increases as the volume fraction of liquid decreases.

C. Densification

After rearrangement, pores remain in the compact if the volume fraction of liquid is low. Additionally, rearrangement can be constricted by solid-solid bonds formed during slow heating. Grain shape accommodation is expected in cases of low liquid contents, leading to pore filling and densification.

Three different transport processes can lead to densification. The first mechanism is termed contact flattening, as first described by Kingery (6). There is a stress at the intergranular contact point due to the capillary force from a wetting liquid. This stress causes preferential dissolution of the solid

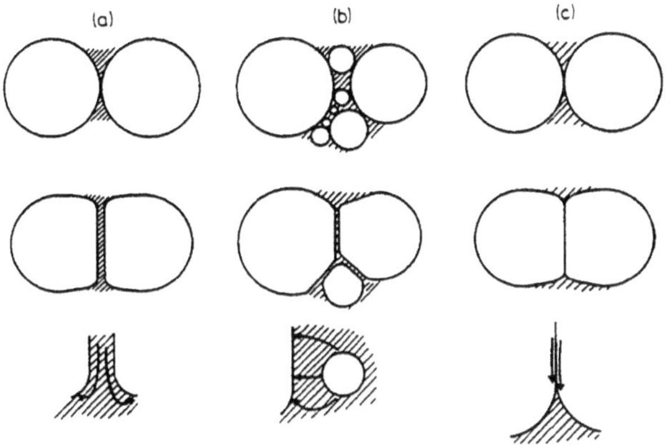

Figure 5.4 The three mechanisms of shape accommodation and neck growth during the intermediate stage; (a) contact flattening, (b) dissolution of fine grains, and (c) solid-state diffusion.

at the contact point with reprecipitation at regions removed from the grain contacts. Densification results from the uniform center-to-center motion of the neighboring grains as illustrated in Figure 5.4a. The flattening of the interface between grains is accomplished by transport of material from the contact with concurrent shrinkage. A second mechanism involves the dissolution of small grains and reprecipitation on large grains (3). In this case, large grains grow and undergo shape accommodation at the expense of neighboring small grains. Again the transport mechanism is solution-reprecipitation as sketched in Figure 5.4b. This second mechanism does not necessarily involve center-to-center approach of the grains. The third mechanism involves growth of the intergrain neck by diffusion through a solid diffusion path as indicated in Figure 5.4c (7,8). The neck growth results in grain shape changes and center-to-center approach of the grains. This form of contact flattening does not involve grain coarsening, but does require a cooperative redistribution process such as described by Swinkels and Ashby (9). Generally, the rates of diffusion through the liquid are so much higher than through the solid that this latter mechanism is not important to the intermediate stage. Exceptions would be for systems with no solid solubility in the liquid such as W-Cu.

The three possible mechanisms of intermediate stage densification are shown in Figure 5.4. Major differences are in the source of material and mode by which densification occurs as contrasted in Table 5.2. All three contribute to better packing and a higher density. Grain growth is an inherent consequence of dissolution of the fine grains, but not of contact flattening. All three mechanisms give shape accommodation and densification. They differ in the transport path, material source, and effect on grain size.

For contact flattening, the rate of material transport from the contact point determines the densification rate. As the size of the contact zone

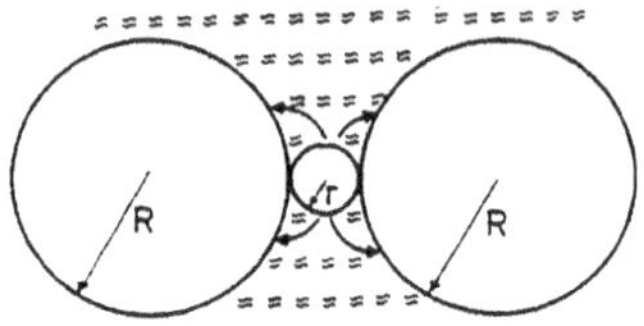

Figure 5.5 Shrinkage by dissolution of a small grain between two large grains.

$$(\Delta L/L_o)^3 = 6\ D\ C\ \gamma\ \Omega\ t/(R^3\ k\ T). \qquad (5.6)$$

Equation (5.6) is similar to Equations (5.1) and (5.3) except for a term of δ/R. Thus, because densification by either mechanism (contact flattening or dissolution of fine grains) depends on solution-reprecipitation, the time dependencies are similar.

There are differing opinions as to the dominance of the possible mechanisms. Whalen and Humenik (23) point out that contact flattening does not explain grain growth and the simultaneous decrease in the number of grains. Often neck growth is not obtained in systems lacking solid solubility in the liquid. Thus, solid state diffusional flow appears to be uncommon (24). Further, when grain growth is inhibited there is no shape accommodation (3,4). These observations support a mechanism where grain shape accommodation takes place by dissolution of the small grains with reprecipitation on the large grains with little center-to-center motion (25). Furthermore, densification correlates with the onset of rapid grain growth in the intermediate stage of liquid phase sintering. Figure 5.6 demonstrates the correlation between dissolution of small grains and densification for the W-Ni system (3). The compact was composed of 48% fine W, 48% coarse W, and 4% Ni. The porosity and percentage of fine tungsten grains are shown versus sintering time. Densification happens in parallel to the elimination of the fine tungsten grains. Generally, the dissolution of small grains and reprecipitation on large grains is supported by microscopic observations during the intermediate stage (1,26-28). However, considering that initial stresses at the contact point are quite large by virtue of its small size, it seems reasonable that contact flattening is initially important (18,29,30).

The difference in densification rates due to contact flattening versus dissolution of small grains with reprecipitation on large grains has been considered by Yoon and Huppmann (3). They define the ratio of volume transport by contact flattening to that due to solution-reprecipitation from small grains to large grains as

$$\xi = \delta/X \qquad (5.7)$$

The presence of an intergranular liquid layer is an important assumption of Kingery's model. For a wetting liquid with zero dihedral angle this assumption is valid. Lange (19) shows that the compressive forces will reduce the thickness of this liquid layer with sintering time to the 1/4 power. However, the liquid layer will always have a finite thickness. Gessinger et al. (7,8) have corrected Kingery's contact stress equation for the dihedral angle contribution, and performed a numerical solution to the contact flattening problem. They found the diffusion controlled shrinkage for two equal sized spheres due to contact flattening was as follows:

$$(\Delta L/L_o)^{3.1} = 12\ \delta\ \Omega\ \gamma\ C\ D\ t/(R^4\ k\ T). \tag{5.3}$$

Equation (5.3) gave the best approximate fit to their numerical solution. Indeed, the shrinkage exponent varied from 3.00 to 3.15 depending on the amount of shrinkage, contact angle, and volume fraction of liquid. Generally, there is a low sensitivity to the amount of liquid and only a slight sensitivity to the other factors.

In the intermediate stage, shrinkage is enhanced by a high solubility of the solid in the liquid. Furthermore, longer times and smaller particle sizes are beneficial to densification. The role of temperature is most pronounced through the diffusivity, although it also has an effect on solubility and surface energy.

Solid-state diffusion is another possible mechanism of densification. As shown in Figure 5.4c, the grain boundary between contacting grains offers a means for neck growth. The predicted neck growth behavior follows Equation (5.3) except the diffusivity is that for the solid-state mechanism. This diffusivity is low in comparison to those typical to diffusion through a liquid. As a consequence, solid-state diffusion is not a significant contributor to densification in most cases of intermediate stage liquid phase sintering.

Modeling of densification by dissolution of small grains with reprecipitation on large grains is a difficult task. The dissolution rate depends on the grain size distribution, packing, and local environment (20,21). Assuming a three grain geometry as shown in Figure 5.5, the net shrinkage by total dissolution of the small grain can be estimated as

$$\Delta L/L_o \approx r/(R + r) \tag{5.4}$$

where the smaller grain has a radius *r*. Note the smaller the center grain, the lower the net shrinkage by dissolution. The rate of dissolution depends on the grain size and its difference from the mean grain size as follows (22):

$$dr/dt = 2\ D\ C\ \gamma\ \Omega\ (r - R)/(k\ T\ r^2\ R) \tag{5.5}$$

with the assumption that the mean grain size is approximately equal to the larger grain radius *R* as shown in Figure 5.5. An approximate solution to Equation (5.5) shows that the particle size will decrease with time to the 1/3 power (20). The shrinkage at any time before the small grain disappears is then approximated as

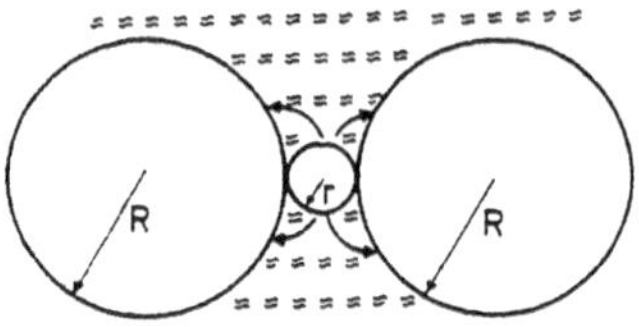

Figure 5.5 Shrinkage by dissolution of a small grain between two large grains.

$$(\Delta L/L_0)^3 = 6\ D\ C\ \gamma\ \Omega\ t/(R^3\ k\ T). \quad (5.6)$$

Equation (5.6) is similar to Equations (5.1) and (5.3) except for a term of δ/R. Thus, because densification by either mechanism (contact flattening or dissolution of fine grains) depends on solution-reprecipitation, the time dependencies are similar.

There are differing opinions as to the dominance of the possible mechanisms. Whalen and Humenik (23) point out that contact flattening does not explain grain growth and the simultaneous decrease in the number of grains. Often neck growth is not obtained in systems lacking solid solubility in the liquid. Thus, solid state diffusional flow appears to be uncommon (24). Further, when grain growth is inhibited there is no shape accommodation (3,4). These observations support a mechanism where grain shape accommodation takes place by dissolution of the small grains with reprecipitation on the large grains with little center-to-center motion (25). Furthermore, densification correlates with the onset of rapid grain growth in the intermediate stage of liquid phase sintering. Figure 5.6 demonstrates the correlation between dissolution of small grains and densification for the W-Ni system (3). The compact was composed of 48% fine W, 48% coarse W, and 4% Ni. The porosity and percentage of fine tungsten grains are shown versus sintering time. Densification happens in parallel to the elimination of the fine tungsten grains. Generally, the dissolution of small grains and reprecipitation on large grains is supported by microscopic observations during the intermediate stage (1,26-28). However, considering that initial stresses at the contact point are quite large by virtue of its small size, it seems reasonable that contact flattening is initially important (18,29,30).

The difference in densification rates due to contact flattening versus dissolution of small grains with reprecipitation on large grains has been considered by Yoon and Huppmann (3). They define the ratio of volume transport by contact flattening to that due to solution-reprecipitation from small grains to large grains as

$$\xi = \delta/X \quad (5.7)$$

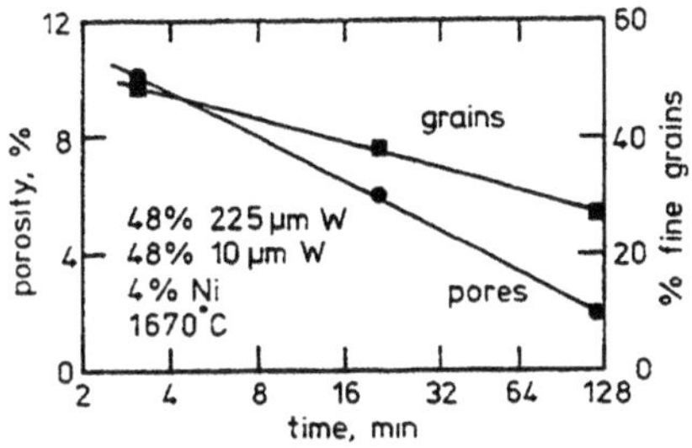

Figure 5.6 A demonstration of densification coupled to dissolution of fine grains in a W-Ni alloy formed from a mixture of coarse and fine tungsten powders (3).

where ξ is the ratio of transport rates, δ is the width of the liquid layer between grains, and X is the radius of the contact zone. They assumed reasonable values for the grain size distribution, diffusion distances, solubilities, and surface energies to arrive at this simple expression for ξ. According to Kingery (6) the liquid width should be a few atomic diameters. Indeed, this is the case in some ceramics with an amorphous (glass) phase between crystalline grains (31-35). For cases of such thin liquid layers, Equation (5.7) indicates that contact flattening is unimportant to densification. For systems such as W-Ni, the thickness is on the order of 1 to 3 μm (17,36). In this case, contact flattening is very significant and dominates initial densification. Thus, contact flattening can account for a considerable portion of the initial shape accommodation process. Only at low volume fractions of liquid and thin boundary layers will contact flattening not be initially dominant as a densification and shape accommodation step. By either process, the densification rate in the intermediate stage will depend on the rate of solution-reprecipitation. As a consequence shrinkage will be proportional to time to the 1/3 power. Figure 5.7 shows two examples of intermediate stage shrinkage data (37,38). The shrinkage data have been plotted on a log-log basis and fit with a line of slope 1/3. Both sets of data agree with the behavior expected for a solution-reprecipitation controlled process.

D. Intergranular Neck Growth

Neck growth between grains occurs during the intermediate stage. Solid necks between contacting grains can be seen in Figure 5.3. Densification is associated with neck growth if the material source is located between the contacting grains. Alternatively, neck growth without densification will occur if the material source is the grain surface away from the point of contact. The growth of a neck between two grains replaces solid-liquid interface area with a grain boundary. This requires a relatively low grain boundary energy. Accordingly, neck growth is the opposite of the liquid penetration observed in the initial stage of liquid phase sintering.

Without shape accommodation, the final neck size X depends on the grain radius R and dihedral angle ϕ as follows:

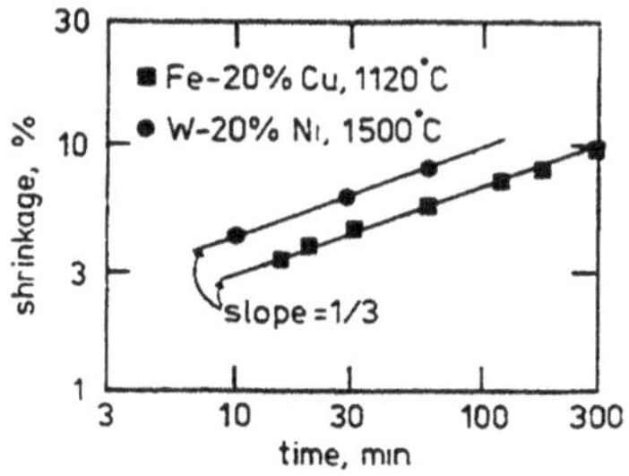

Figure 5.7 Shrinkage during solution-reprecipitation for two alloy systems, W-Ni (37) and Fe-Cu (38). The behavior follows the predicted one-third power relation.

$$X = R \sin(\phi/2). \tag{5.8}$$

Since the dihedral angle depends on the ratio of solid-solid to solid-liquid surface energies, then the final neck size ratio can be calculated from these energies. Figure 2.31 shows the resulting relation between X/R and the energy ratio. Various two-grain geometries are shown in Figure 2.31, corresponding to selected dihedral angles. For unequal spheres, the final neck size is determined by the ratio of the grain sizes and the dihedral angle. Figure 5.8 plots the final neck size ratio X/R against the grain size ratio (small divided by large) for various dihedral angles at 15° intervals. The neck radius divided by the radius of the larger grain decreases as the grains differ in size. Alternatively, the neck size as normalized by the smaller grain radius increases with the difference in grain sizes. At high volume fractions of solid, there is an energy minimization possible through larger neck sizes. White (22) has shown a slight increase in the neck size ratio as the volume fraction of liquid decreases in Mg-Ca-Si-O systems. Figure 5.9 plots his result for the periclase-silicate system at a constant dihedral angle. As the amount of silicate increases, the solid content and neck size ratio both decrease. Park and Yoon (39) predict such behavior for low dihedral angles and low volume fractions of liquid. An example of the predicted behavior for a dihedral angle of 60° is shown in Figure 5.10. Although the underlying model has several simplifying assumptions, it provides an instructive demonstration of shape accommodation effects on final neck size.

Initially, there is no neck between contacting grains. At the end of the intermediate stage there is a stable neck as determined by the dihedral angle, grain size, and volume fraction of solid. Thus, during the intermediate stage, neck growth occurs by mass transport from the grain contact point or grain surface. Gessinger et al. (7,8) solved the neck growth rate problem assuming transport from the contact point. Accompanying neck growth by this transport path is densification. They found the amount of liquid does not significantly change the neck growth rate as long as there is sufficient liquid to cover the neck. The contact angle has a slight effect on the growth rate. The general result of their calculations is a neck size equation as follows:

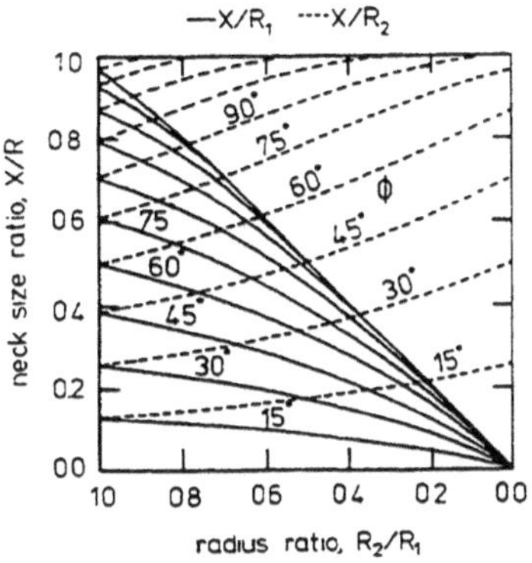

Figure 5.8 The neck size ratio for contacting grains with differing sizes and several possible dihedral angles. The neck size ratio is shown based on the radius of each grain.

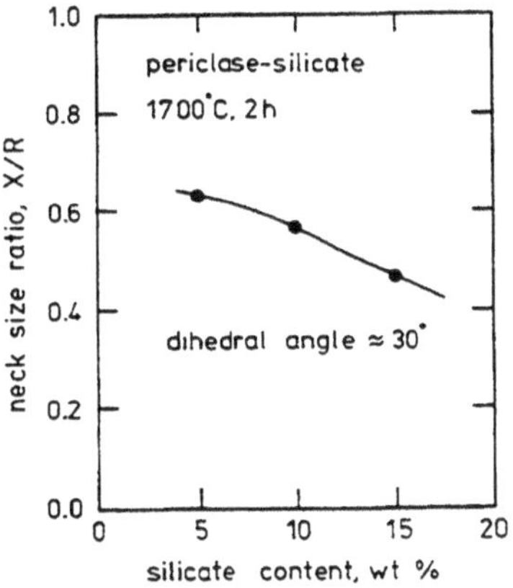

Figure 5.9 The neck size ratio as a function of silicate content in a magnesia-silica-calcia system. The neck size ratio enlarges as the amount of liquid decreases at a constant dihedral angle (22).

$$(X/R)^n = K\,t \tag{5.9}$$

where t is the isothermal time and K is described below. The exponent n was found to range from 6 to 7, with a best approximation of 6.22. The factor K in Equation (5.9) is a collection of terms as follows:

$$K = 96\,\delta\,D\,\gamma\,\Omega/(R^4\,k\,T) \tag{5.10}$$

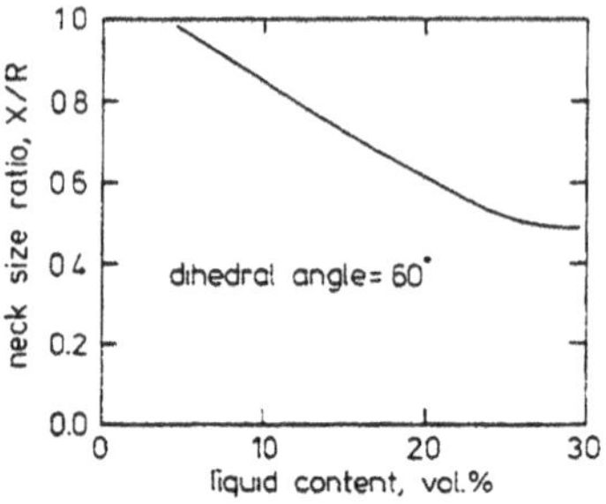

Figure 5.10 The predicted enlargement of the neck size ratio as the volume fraction of liquid decreases for a constant dihedral angle of 60° (39).

with δ equal to the width of the grain boundary diffusion zone, D equal to the diffusivity, γ equal to the liquid-vapor surface energy, Ω equal to the atomic volume, k equal to Boltzmann's constant, and T equal to the absolute temperature. Equation (5.9) is similar to that predicted by Kingery (6) for shrinkage through contact flattening. Kingery also assumed the liquid-vapor surface energy provided the contact stress, but mass flow was along a liquid rich layer located at the intergranular boundary. For the situation described by Kingery, K of Equation (5.10) should be multiplied by the solid solubility in the liquid.

As an alternative, Courtney (40) and Kaysser et al. (41) have considered neck growth by solution-reprecipitation through the liquid, with the mass source from the grain surface away from the contact point. These treatments assume a large quantity of liquid; thus, there is no effect from the volume fraction of liquid, contact angle, nor liquid-vapor surface energy. Furthermore, these treatments assume no solid-solid interfacial energy, which corresponds to a dihedral angle of 180°. A symmetric flux field around the contacting grains leads to a predicted neck growth equation as follows:

$$(X/R)^5 = 20\ D\ C\ \gamma\ \Omega\ t/(R^3\ k\ T) \tag{5.11}$$

for the early portion of neck growth, and at long times,

$$(X/R)^6 = 18\ D\ C\ \gamma\ \Omega\ t/(\pi\ R^3\ k\ T). \tag{5.12}$$

Equations (5.11) and (5.12) describe neck growth by solution-reprecipitation without densification. One prediction from the modeling by Courtney is that the neck growth rate should be slower as the volume fraction of solid increases.

Equations (5.9) through (5.12) describe neck growth in terms of only one mechanism. Actually, neck growth occurs by several simultaneous mechanisms involving coalescence, solid state diffusion, and solution-reprecipitation. Other problems with the past treatments are the geometric simplifications of uniform spheres, equal sized grains, and no dihedral angle.

Neck growth is limited by the stable size dictated by energy minimization (39). Once a stable neck size ratio (X/R) is obtained in the intermediate stage, further neck growth is dependent on grain growth. The neck size ratio will remain constant during grain growth; thus, the actual neck radius will grow with time to the 1/3 power (1,42,43). After obtaining the equilibrium neck size ratio as set by the dihedral angle, the number of necks per grain will remain constant, but the total number of necks per unit volume will vary with inverse time as shown in Figure 5.1. Densification occurs during neck growth by contact flattening, involving mass transport from the grain contact point to the neck. The diffusion responsible for neck growth occurs in either a solid grain boundary or in a liquid coated boundary. The neck size will vary with time to approximately the 1/6 power, while shrinkage will vary with time to approximately the 1/3 power.

E. Coalescence

After liquid flow, grains are put into contact by a wetting liquid. Coalescence is a possible intermediate stage densification and coarsening mechanism involving the contacting grains. A sketch of coalescence is given in Figure 5.11. Contacting grains of dissimilar size fuse into a single grain by a continuous process of directional grain growth and grain reshaping. The reshaping process is most typically by concurrent solution-reprecipitation. The grain size increases by coalescence and the number of grains continuously decreases.

Grain coalescence early in liquid phase sintering has been reported for several materials, including Fe-Cu (1,41,42,44,45), tungsten-based alloys (17,46-52), cemented carbides (53), Mo-Ni-Fe (28), and Cu-Ag (42,54). For most systems, contacts form quickly when the melt appears. There is a finite probability that the contact will form a low-angle grain boundary. A low misorientation angle between the contacting grains results in a high probability of subsequent coalescence (55,56). Alternatively, for grains with differing sizes, the driving force for coalescence is boundary migration due to curvature. Other possible causes of coalescence are chemical, strain, or temperature gradients (28,47,48,57).

Contact formation is a precursor to neck growth and coalescence. New contacts between grains are induced by gravity, thermal motion, and settling. Many contacts exist prior to liquid formation, especially with high volume fractions of solid and high green densities. Courtney (58) has given a detailed analysis of contact formation processes for low volume fractions of solid.

In crystalline powder systems, grain boundaries form between grains after a short sintering time if the dihedral angle is greater than zero. Accordingly, the connectivity and contiguity increase. For volume fractions of solid below approximately 35%, there will be rapid grain motion in the liquid. During motion, grain contacts can form which can possibly coalesce. An isolated microstructure is anticipated when the time for coalescence is short compared to the time between contacts. Alternatively, a skeletal

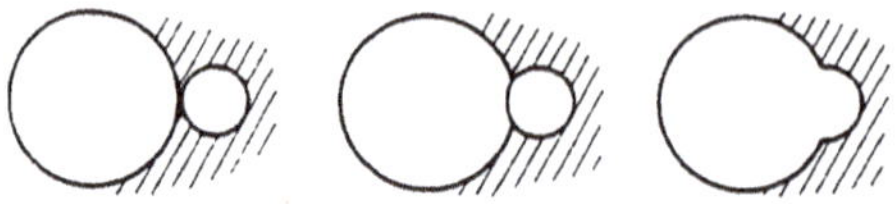

Figure 5.11 A schematic diagram of the steps leading to grain growth by coalescence.

microstructure occurs when the coalescence time is long compared to the frequency of grain contacts. This latter case is most typical to liquid phase sintering where the volume fraction of solid exceeds 70%. Figure 5.12 contrasts two extreme examples of the microstructure for a 50 volume percent solid system. The isolated structure would be favored by a low dihedral angle or rapid coalescence as compared to the contact frequency. A skeletal structure is more typical to liquid phase sintered materials, where the coalescence time is long and there is a finite dihedral angle. Since the packing coordination of a grain increases with the volume fraction of solid, the probability of coalescence is increased by a high volume fraction of solid (59,60). However, the grain motion in the liquid is inhibited by a high volume fraction of solid. Consequently, coalescence is estimated to be important only for the first 10 minutes after liquid formation for most systems (46,53,61).

Once a contact has formed, the driving force for coalescence is a lowering of the system energy by elimination of an interface. Three transport paths can act to coalesce contacting grains; grain boundary migration by solid state diffusion, grain boundary migration by diffusion across an intermediate thin liquid layer, or solution-reprecipitation from the small to the large grain. Figure 5.13 compares these three possible transport paths. There is evidence for coalescence by grain boundary migration by either liquid or solid diffusion events (17,48-50,55,62). Coalescence by solution-reprecipitation as an analog to evaporation-condensation has been modeled but not identified with any experimental data (40,41,58). This latter concept of coalescence applies only to amorphous solids which do not have a solid-solid grain boundary. Generally, the increased energy to enlarge the grain boundary area during migration requires simultaneous solution-reprecipitation.

Coalescence is easiest when the grain boundary between contacting grains has a low misorientation, but this is a low probability situation. A low misorientation angle corresponds to a low grain boundary energy and a high dihedral angle. The probability of coalescence due to a low misorientation angle is estimated as 0.01 or less (55,62). Accordingly, low misorientation grain boundaries are a minor contributor to grain coarsening in liquid phase sintering.

Another observed coalescence process is by liquid boundary migration. A difference in strain between contacting grains nucleates the growth process. After the process starts, the driving force is a change in the chemistry across the moving boundary, with a preferred crystallographic

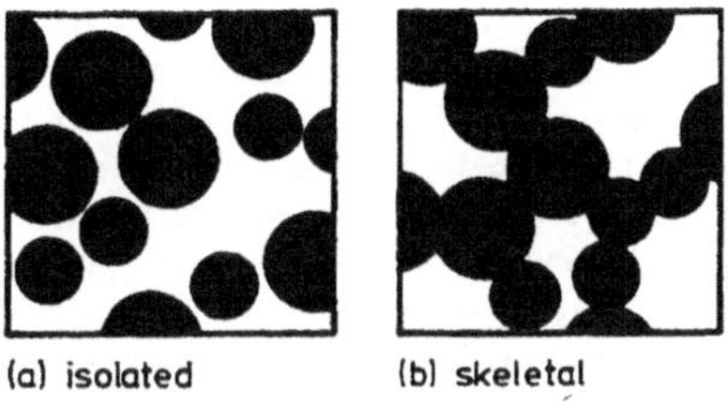

Figure 5.12 The contrast between an isolated and skeletal microstructure for 50 vol.% liquid. A skeletal structure is expected when coalescence is slow (58).

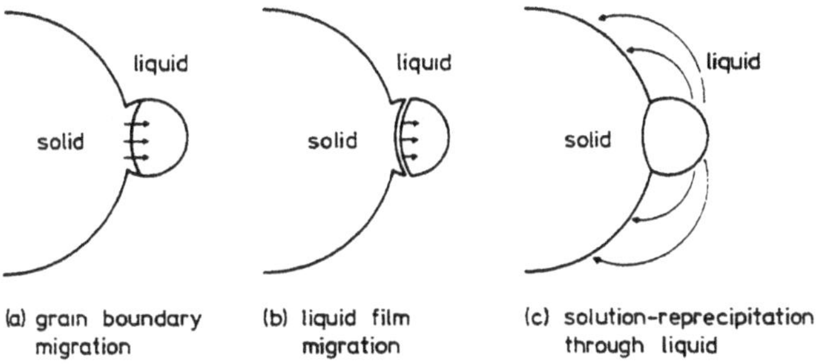

Figure 5.13 Three possible mechanisms of coalescence between contacting grains; (a) solid-state grain boundary motion, (b) liquid film migration, and (c) solution reprecipitation through the liquid.

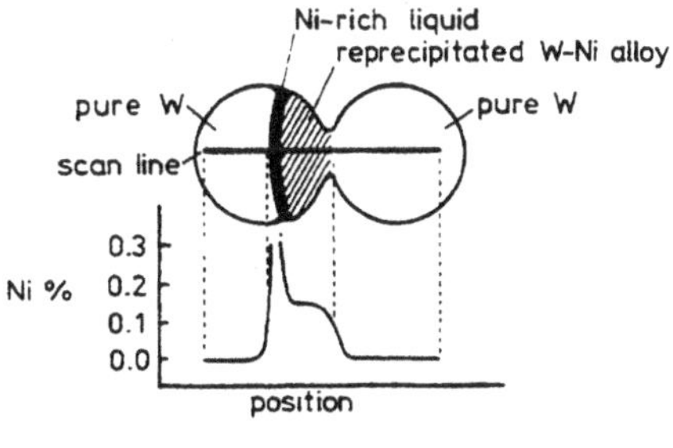

Figure 5.14 Coalescence by the motion of a liquid film through a contacting grain for W-Ni. The concentration of Ni versus position along the scan line is shown below the two grains (2).

orientation for growth (63). The structure of two tungsten grains coalescing by liquid boundary migration is shown in Figure 5.14, with the microprobe determined concentration profile shown below the grains (2). The boundary motion is by solution-reprecipitation across the boundary, where the dissolving grain is a pure substance and the reprecipitated material is an alloy. The boundary curvature is the opposite from that expected for solid-state grain growth. This indicates a chemical driving force for liquid boundary motion controlled by solution-reprecipitation as discussed by Jones (64). Other factors such as strain in the liquid layer can affect the driving force for boundary motion, but the chemical concentration gradient appears to be dominant (36). At the maximum observed growth rate, the boundary moves as fast as 20 μm/min. Thus far, liquid boundary migration as a coalescence process has been demonstrated only for W-Ni alloys (17,48-50,65).

The rate of boundary motion between two coalescing grains depends on the chemical potential difference. For grain boundary motion by solid-state diffusion, the driving force is the difference in grain sizes. Two contacting grains with a difference in radii and a nonzero dihedral angle are shown in Figure 5.15. The size of the neck at equilibrium will be dependent on the dihedral angle, while the radius of curvature of the grain boundary r is given in terms of the two grain radii as

$$r = (\gamma_{SS}/\gamma_{SL})\ [R_1\ R_2/(R_1 - R_2)] \tag{5.13}$$

where the interfacial energy ratio can be expressed in terms of the dihedral angle. The difference in energy between the two grains varies with the inverse of the boundary curvature. Figure 5.16 shows the normalized boundary curvature versus the normalized grain radius ratio (66). As the two grains differ in size, the boundary becomes more curved and less stable. Diffusion across the boundary results in a more stable structure. The coarsening rate will be faster as the two grains differ in size. At high dihedral angles and large grain size differences, there will be no energy barrier to boundary motion. Thus, low misorientation angles between the grains, high dihedral angles, and large grain size differences contribute to easy coalescence. For low dihedral angles, the motion of the boundary through the small grain will increase the system energy since the area of the grain boundary will increase. Thus, there is an energy barrier impeding coalescence. However, the simultaneous action of solution-reprecipitation from the small to large grain can act to eliminate the energy barrier. In this case, the coalescence rate will be dependent on the rate of solution-reprecipitation (20,44,55,56), and the coarsening time will vary with the cube of the smaller grain size (42,45).

Coalescence is favored by a high volume fraction of solid, high diffusivity, and high dihedral angle. In systems like W-Cu, W-Au, and WC-Cu, where there is no significant solid solubility in the liquid, observations of grain growth can only be explained by coalescence (51,53). Because coalescence is dependent on grain contacts with large size differences or low crystallographic misorientations, it is most active in the early portion of the intermediate stage. Later the microstructure stabilizes and a rigid skeleton inhibits new contacts. Measurements on W-Ni-Fe heavy alloys suggest that coalescence decays in importance rapidly as sintering progresses (46). Figure 5.17 plots the measured percentage of solid grain pairs undergoing

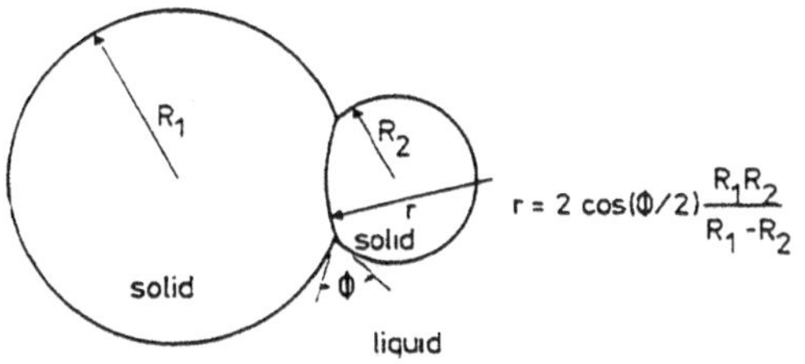

Figure 5.15 The equilibrium geometry of two contacting grains of differing sizes. The intergranular grain boundary curvature depends on the grain sizes and the dihedral angle.

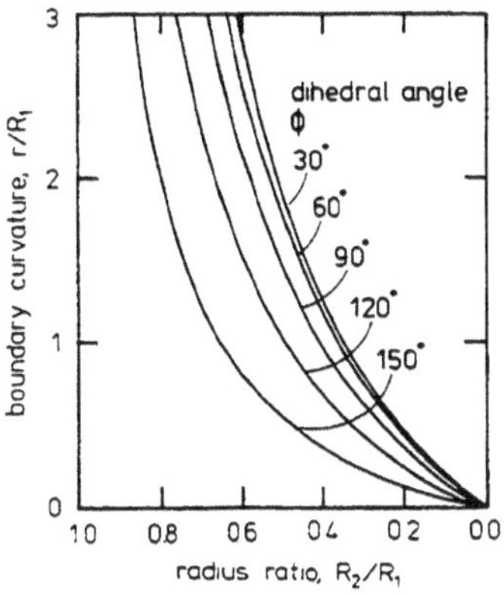

Figure 5.16 The normalized equilibrium grain boundary curvature shown as a function of the dihedral angle and the ratio of the two contacting grains.

coalescence during liquid phase sintering. After one hour of sintering, coalescence is contributing to coarsening at only 1% of the contacts. Warren and Waldron (67) analyzed grain coarsening rates versus contiguity for cemented carbides in 20 volume percent liquid. As shown in Figure 5.18, the rate of grain growth decreased as the contiguity increased. They contend such behavior directly contradicts the coalescence models. An increased rate of grain growth is expected as coalescence becomes more important. With a high contiguity there is more grain contact and more opportunity for coalescence, but less surface area over which solution-reprecipitation can occur. This later factor seems to dominate coarsening of cemented carbides. Finally, Kaysser et al. (41) show that the very largest grains are most typically involved in coalescence. Since these grains constitute a small fraction of the grain size distribution, it is again concluded coalescence has a small probability.

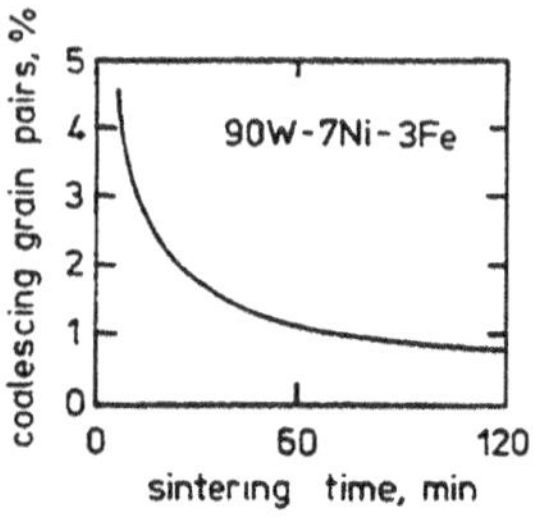

Figure 5.17 The percentage of contacting grains undergoing coalescence versus sintering time for a W-Ni-Fe alloy, showing coalescence decreases in importance with prolonged sintering (46).

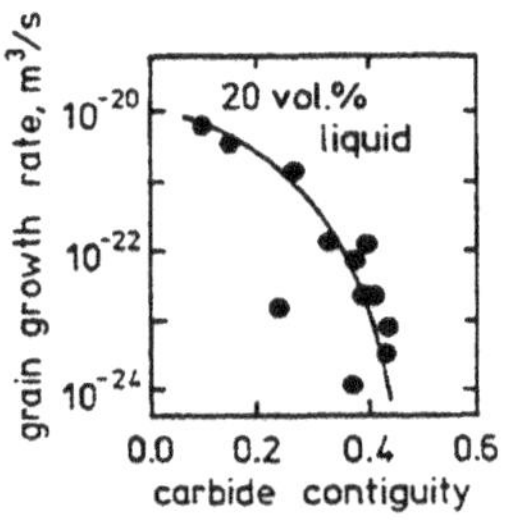

Figure 5.18 The grain growth rate versus carbide contiguity for various carbides in a 20 vol.% cobalt-based liquid. An increase in the growth rate is expected if coalescence is dominant (67).

In summary, coalescence is an often observed mechanism of densification and microstructural coarsening in the early portion of liquid phase sintering. Coalescence contributes to a decrease in the number of grains and an increase in the mean grain size. It can occur by solid-state diffusion across grain boundaries, solution-reprecipitation through the liquid, or by the migration of a liquid layer through the solid. Solid-state diffusion is the most frequently observed mechanism; however, in most cases of coalescence a cooperative solution-reprecipitation process is necessary. The data and analysis suggest that a small fraction (approximately 3%) of the initial grain contacts will undergo coalescence in the first ten minutes of liquid phase sintering. Coalescence requires an increase in grain boundary area for most practical situations and is dependent on simultaneous solution-reprecipitation. A large solid volume fraction and a high dihedral angle aid the coalescence process. A low misorientation angle between contacting grains provides the needed high dihedral angle.

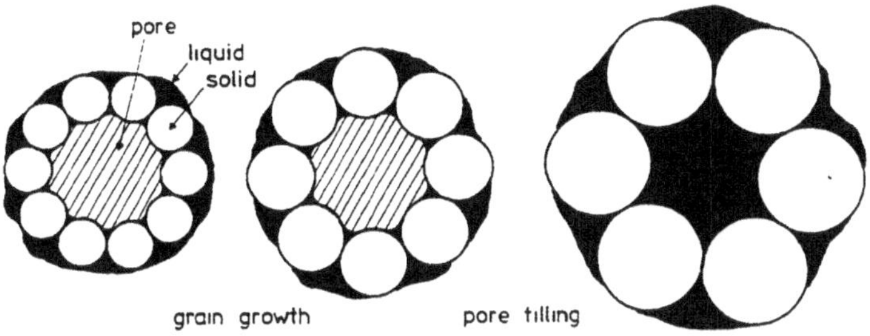

Figure 5.19 Pore filling during grain growth. A large pore is stable until grain growth increases the liquid meniscus radius sufficiently for capillary refilling of the pore.

F. Pore Filling

Grain shape accommodation initially contributes to pore filling in the intermediate stage of liquid phase sintering. However, pores formed at prior additive particle sites are often stable through a considerable portion of the intermediate stage. The few remaining large pores are difficult to fill.

Pore filling requires grain growth and shape accommodation. Kwon and Yoon (68) report pore filling has a radial component wherein full density occurs at the compact center and spreads outward during the intermediate stage. Kim et al. (69) found the radial flow of liquid only occurred when the sintered density was near 80% of theoretical. The higher the contact angle, the greater the difficulty in removing the residual pores. A process of liquid repenetration occurs during grain growth as illustrated in Figure 5.19. A large pore surrounded by grains remains unfilled because of capillary wetting of the intergrain cavities. During prolonged sintering the grains grow and eventually the liquid reaches a favorable condition for refilling the pore (70,71). The liquid meniscus radius at the pore-liquid-grain contact is given as r_m, as shown in Figure 5.20,

$$r_m = R\ (1 - \cos\alpha)/\cos\alpha \tag{5.14}$$

where R is the grain radius and α is the angle from the grain center to the solid-liquid-vapor contact point. During grain growth the grain radius and the meniscus radius will increase. Eventually, the capillary pressure associated with the meniscus will favor liquid flow into the pore since this will give a lower liquid pressure. For a zero contact angle, pore filling will occur when the pore radius and the meniscus radius are equal. If the contact angle is greater than zero, the meniscus radius must exceed the grain radius for pore filling. Thus, low contact angles are beneficial to pore filling. Kaysser and Petzow (72) suggest that pore filling is difficult for contact angles over 22°. The sequence of pore filling steps is shown schematically in Figure 5.19 and micrographs of refilled pores are given in Figures 5.21 and 5.22. The radius of a stable pore varies with the inverse of the grain size (4). Since grain growth follows a cube-root dependence on time, the final densification

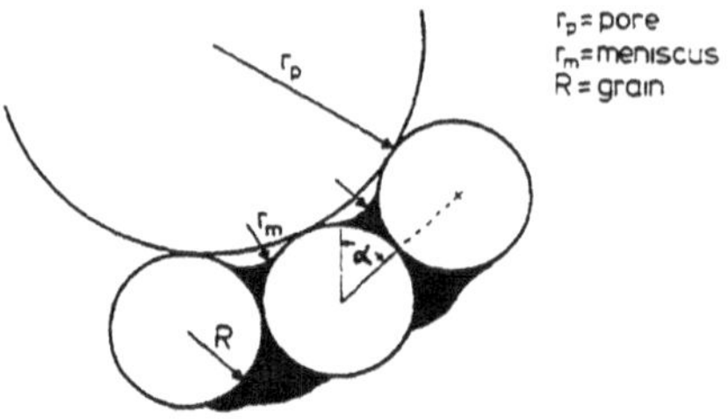

Figure 5.20 The calculation model for pore refilling based on spherical grains surrounding the pore. Pore refilling is dependent on the liquid meniscus radius exceeding the pore radius.

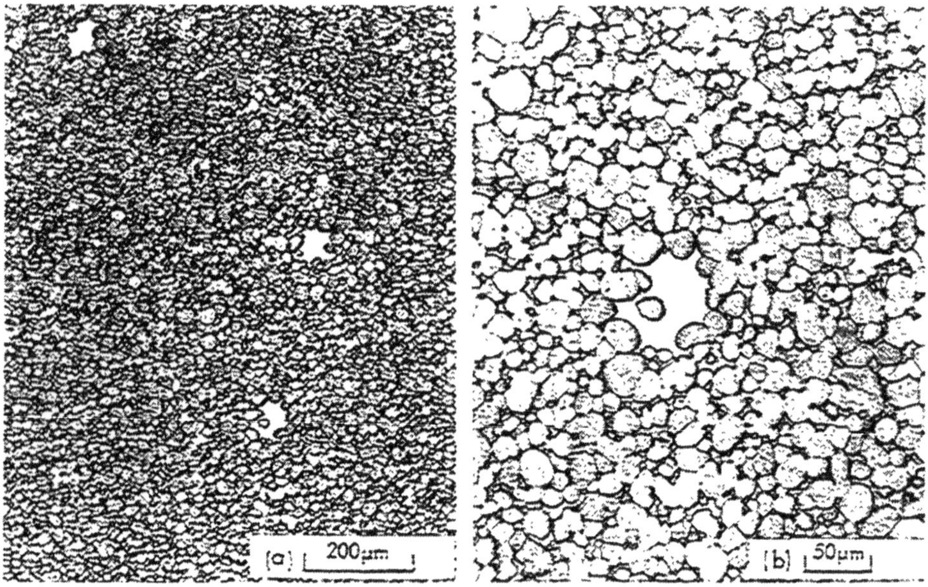

Figure 5.21 Optical micrographs of refilled pores in a tungsten heavy alloy after liquid phase sintering (photos courtesy of E. G. Zukas).

will also be dependent on time to the 1/3 power. If there is trapped gas in the pore, then it will be partially refilled. After pore filling, solid grains will eventually reform in the liquid and eliminate any microstructural inhomogeneities. The coarser the additive particle size, the longer it takes to eliminate the residual pores. This relation explains the often observed benefit of small additive particle sizes.

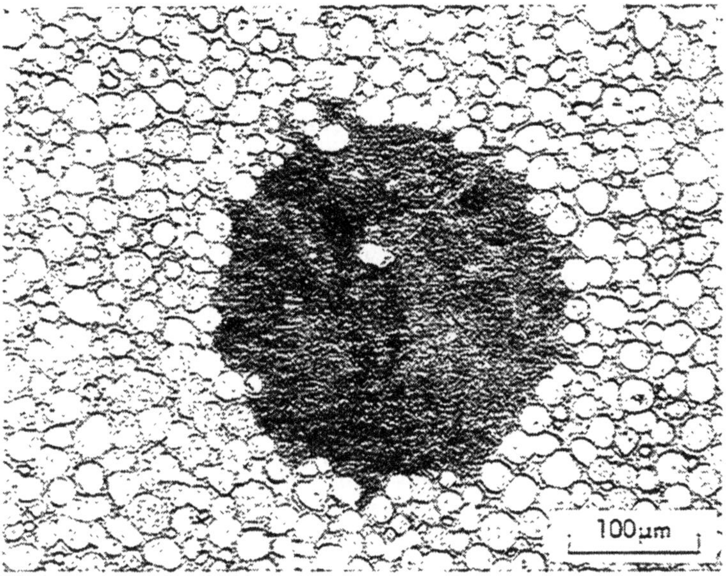

Figure 5.22 A refilled pore in a 90W-7Ni-3Fe heavy alloy after sintering at 1470°C for 30 minutes showing a dislodged grain in the liquid (photo courtesy of B. H. Rabin).

G. Summary

The intermediate stage of liquid phase sintering is characterized by densification, grain growth, grain shape accommodation, contact flattening, pore elimination, coalescence, and neck growth. Dissolution and reprecipitation processes dominate microstructural development during the intermediate stage. Densification is associated with grain shape accommodation, which takes place by contact flattening at grain contacts, dissolution of small grains with reprecipitation on large grains, and coalescence involving grain boundary migration and cooperative solution-reprecipitation. The controlling step for each of these mechanisms is typically diffusion through the liquid, although interfacial reaction control is occasionally observed.

Grain shape accommodation is important to densification in systems with low volume fractions of liquid. The grains deviate in shape from a low surface area geometry to allow a higher packing density and provide liquid for pore filling. The deviation from a minimum energy shape gives a net energy decrease by eliminating residual porosity and the associated surface energy. Both solubility of the solid in the liquid and high diffusion rates are beneficial for densification.

Coalescence of contacting grains occurs during the early portion of the intermediate stage; however, most data indicate this is a transient contributor to microstructural coarsening. Coalescence is rapid in those cases where the misorientation angle between contacting grains is small. Alternatively, with large misorientation angles, a high volume fraction of solid and a high dihedral angle aid coalescence. Boundary migration, either as a solid or liquid boundary, occurs by diffusion. A chemical potential difference or grain size difference provides the driving force for migration. However, there is an energy barrier associated with direct boundary migration through the smaller grain. As a consequence, a cooperative solution-reprecipitation process is necessary.

Neck growth occurs between contacting grains when the dihedral angle is larger than zero. The mechanisms of neck growth are similar to those giving shape accommodation and densification. A stable neck size is established by the dihedral angle, volume fraction of solid, and grain size ratio. Mass flow to attain the stable neck size ratio is through solution-reprecipitation or grain boundary diffusion.

Pore filling is the final action associated with the intermediate stage. Typically, the largest pores remain stable to late in the sintering cycle. The elimination of the largest pores requires grain growth and grain shape accommodation. Grain growth changes the liquid meniscus size at the solid-liquid-vapor interface, eventually allowing capillary refilling of the pores. Thus, final pore elimination will depend on the rate of grain growth. In turn, most of these events are dependent on the rate of mass transport through the liquid, which justifies the emphasis on solution-reprecipitation during the intermediate stage.

In the next chapter the concurrent microstructural coarsening events such as grain growth are discussed. It is emphasized that coarsening is also via solution-reprecipitation and is separated here for convenience.

H. References

1. R. Watanabe and Y. Masuda, "The Growth of Solid Particles in Fe-20 wt.% Cu Alloy During Sintering in the Presence of a Liquid Phase," *Trans. Japan Inst. Met.*, 1973, vol.14, pp.320-326.
2. W. J. Huppmann, "The Elementary Mechanisms of Liquid Phase Sintering. 2. Solution - Reprecipitation," *Z. Metallkde.*, 1979, vol.70, pp.792-797.
3. D. N. Yoon and W. J. Huppmann, "Grain Growth and Densification During Liquid Phase Sintering of W-Ni," *Acta Met.*, 1979, vol.27, pp.693-698.
4. W. A. Kaysser, O. J. Kwon, and G. Petzow, "Pore Formation and Pore Elimination During Liquid Phase Sintering," *Proceedings P/M-82*, Associazione Italiana di Metallurgia, Milano, Italy, 1982, pp.23-30.
5. H. H. Park, S. J. Cho, and D. N. Yoon, "Pore Filling Process in Liquid Phase Sintering," *Metall. Trans. A*, 1984, vol.15A, pp.1075-1080.
6. W. D. Kingery, "Densification During Sintering in the Presence of a Liquid Phase. 1. Theory," *J. Appl. Phys.*, 1959, vol.30, pp.301-306.
7. G. H. Gessinger and H. F. Fischmeister, "A Modified Model for the Sintering of Tungsten with Nickel Additions," *J. Less-Common Metals*, 1972, vol.27, pp.129-141.
8. G. H. Gessinger, H. F. Fischmeister, and H. L. Lukas, "A Model for Second-Stage Liquid-Phase Sintering with a Partially Wetting Liquid," *Acta Met.*, 1973, vol.21, pp.715-724.
9. F. B. Swinkels and M. F. Ashby, "Role of Surface Redistribution in

Sintering by Grain Boundary Transport," *Powder Met.*, 1980, vol.23, pp.1-7.
10. L. Lindau and K. G. Stjernberg, "Grain Growth in TiC-Ni-Mo and TiC-Ni-W Cemented Carbides," *Powder Met.*, 1976, vol.19, pp.210-213.
11. S. Sarian and H. W. Weart, "Kinetics of Coarsening of Spherical Particles in a Liquid Matrix," *J. Appl. Phys.*, 1966, vol.37, pp.1675-1681.
12. H. E. Exner, E. Santa Marta, and G. Petzow, "Grain Growth in Liquid-Phase Sintering of Carbides," *Modern Developments in Powder Metallurgy*, vol.4, H. H. Hausner (ed.), Plenum Press, New York, NY, 1971, pp.315-325.
13. H. E. Exner, "Ostwald-Reifung von Ubergangsmetallkarbiden in Flussingem Nickel und Kobalt," *Z. Metallkde.*, 1973, vol.64, pp.273-279.
14. G. C. Kuczynski and O. P. Gupta, "Model Experiments of Sintering in the Presence of Liquid Phase," *Sintering Theory and Practice*, M. M. Ristic (ed.), International Team for Studying Sintering, Beograd, Yugoslavia, 1973, pp.187-200.
15. A. L. Prill, H. W. Hayden, and J. H. Brophy, "A Reanalysis of Data on the Solution-Reprecipitation Stage of Liquid-Phase Sintering," *Trans. TMS-AIME*, 1965, vol.233, pp.960-964.
16. W. J. Huppmann, "Sintering in the Presence of a Liquid Phase," *Sintering and Catalysis*, G. C. Kuczynski (ed.), Plenum Press, New York, NY, 1975, pp.359-378.
17. Z. S. Nikolic and W. J. Huppmann, "Computer Simulation of Chemically Driven Grain Growth During Liquid Phase Sintering," *Acta Met.*, 1980, vol.28, pp.475-479.
18. P. V. Hobbs and B. J. Mason, "The Sintering and Adhesion of Ice," *Phil. Mag.*, 1964, vol.9, pp.181-197.
19. F. F. Lange, "Liquid Phase Sintering: Are Liquids Squeezed Out from Between Compressed Particles," *J. Amer. Ceramic Soc.*, 1982, vol.65, p.C23.
20. S. Takajo, W. A. Kaysser, and G. Petzow, "Analysis of Particle Growth by Coalescence During Liquid Phase Sintering," *Acta Met.*, 1984, vol.32, pp.107-113.
21. P. W. Voorhees, "Ostwald Ripening in Two Phase Mixtures", Ph.D. Thesis, Rensselaer Polytechnic Institute, Troy, NY, 1982.
22. J. White, "Microstructure and Grain Growth in Ceramics in the Presence of a Liquid Phase," *Sintering and Related Phenomena*, G. C. Kuczynski (ed.), Plenum Press, New York, NY, 1973, pp.81-108.
23. T. J. Whalen and M. Humenik, "Sintering in the Presence of a Liquid Phase," *Sintering and Related Phenomena*, G. C. Kuczynski, N. Hooton, and C. Gibbon (eds.), Gordon and Breach, New York, NY, 1967, pp.715-74.
24. G. H. Gessinger, H. F. Fischmeister, and H. L. Lukas, "The Influence of a Partially Wetting Second Phase on the Sintering of Solid Particles," *Powder Met.*, 1973, vol.16, pp.119-127.
25. W. A. Kaysser, M. Zivkovic, and G. Petzow, "Shape Accommodation During Grain Growth in the Presence of a Liquid Phase," *J. Mater. Sci.*, 1985, vol.20, pp.578-584.
26. S. S. Kim and D. N. Yoon, "Formation of Etch Boundaries at the Interface of Mo Grains and a Liquid Ni-Fe Matrix During Cyclic Annealing," *Metallog.*, 1983, vol.16, pp.249-253.
27. W. A. Kaysser and G. Petzow, "Recent Conceptions on Liquid Phase Sintering," *Proceedings Sintering Theory and Practice Conference*, The

Metals Society, London, UK, 1984, pp.10.1-10.6.
28. S. S. Kim and D. N. Yoon, "Coarsening Behaviour of Mo Grains Dispersed in Liquid Matrix," *Acta Met.*, 1983, vol.31, pp.1151-1157.
29. W. D. Kingery, "Sintering in the Presence of a Liquid Phase," *Kinetics of High-Temperature Processes*, W. D. Kingery (ed.), John Wiley, New York, NY, 1959, pp.187-194.
30. V. N. Eremenko, Y. V. Naidich, and I. A. Lavrinenko, *Liquid Phase Sintering*, Consultants Bureau, New York, NY, 1970.
31. L. K. V. Lou, T. E. Mitchell, and A. H. Heuer, "Impurity Phases in Hot Pressed Silicon Nitride," *J. Amer. Ceramic Soc.*, 1984, vol.67, pp.392-396.
32. D. R. Clarke and G. Thomas, "Microstructure of Yttria Fluxed Hot Pressed Silicon Nitride," *J. Amer. Ceramic Soc.*, 1978, vol.61, pp.114-118.
33. D. R. Clarke, N. J. Zaluzec, and R. W. Carpenter, "The Intergranular Phase in Hot Pressed Silicon Nitride: I, Elemental Composition," *J. Amer. Ceramic Soc.*, 1981, vol.64, pp.601-607.
34. R. L. Tsai and R. Raj, "Creep Fracture in Ceramics Containing Small Amounts of a Liquid Phase," *Acta Met.*, 1982, vol.30, pp.1043-1058.
35. R. Raj, "Morphology and Stability of the Glass Phase in Glass-Ceramic Systems," *J. Amer. Ceramic Soc.*, 1981, vol.64, pp.245-248.
36. Y. D. Song and D. N. Yoon, "The Driving Force for Chemically Induced Migration of Molten Ni Films Between W Grains," *Metall. Trans. A*, 1984, vol.15A, pp.1503-1505.
37. W. J. Huppmann and H. Riegger, "Liquid Phase Sintering of the Model System W-Ni," *Inter. J. Powder Met. Powder Tech.*, 1977, vol.13, pp.243-247.
38. W. D. Kingery and M. D. Narasimhan, "Densification During Sintering in the Presence of a Liquid Phase 2. Experimental," *J. Appl. Phys.*, 1959, vol.30, pp.307-310.
39. H. H. Park and D. N. Yoon, "Effect of Dihedral Angle on the Morphology of Grains in a Matrix Phase," *Metall. Trans. A*, 1985, vol.16A, pp.923-928.
40. T. H. Courtney, "A Reanalysis of the Kinetics of Neck Growth During Liquid Phase Sintering," *Metall. Trans. A*, 1977, vol.8A, pp.671-677.
41. W. A. Kaysser, S. Takajo, and G. Petzow, "Particle Growth by Coalescence During Liquid Phase Sintering of Fe-Cu," *Acta Met.*, 1984, vol.32, pp.115-122.
42. Y. Masuda and R. Watanabe, "Ostwald Ripening Processes in the Sintering of Metal Powders," *Sintering Processes*, G. C. Kuczynski (ed.), Plenum Press, New York, NY, 1980, pp.3-21.
43. R. Watanabe and Y. Masuda, "The Growth of Solid Particles in Some Two-Phase Alloys During Sintering in the Presence of a Liquid Phase," *Sintering and Catalysis*, G. C. Kuczynski (ed.), Plenum Press, New York, NY, 1975, pp.389-398.
44. A. N. Niemi and T. H. Courtney, "Microstructural Development and Evolution in Liquid-Phase Sintered Fe-Cu Alloys," *J. Mater. Sci.*, 1981, vol.16, pp.226-236.
45. W. A. Kaysser, S. Takajo, and G. Petzow, "Particle Growth by Coalescence During Liquid Phase Sintering of Fe-Cu," *Sintering - Theory and Practice*, D. Kolar, S. Pejovnik, and M. M. Ristic (eds.), Elsevier Scientific, Amsterdam, Netherlands, 1982, pp.321-327.
46. R. V. Makarova, O. K. Teodorovich, and I. N. Frantsevich, "The Coalescence Phenomenon in Liquid-Phase Sintering in the Systems

Tungsten-Nickel-Iron and Tungsten-Nickel-Copper," *Soviet Powder Met. Metal Ceram.*, 1965, vol.4, pp.554-559.
47. E. G. Zukas, P. S. Z. Rogers, and R. S. Rogers, "Unusual Spheroid Behavior During Liquid Phase Sintering," *Inter. J. Powder Met. Powder Tech.*, 1977, vol.13, pp.27-38.
48. L. Kozma, W. J. Huppmann, L. Bartha, and P. Mezei, "Initiation of Directional Grain Growth During Liquid-Phase Sintering of Tungsten and Nickel," *Powder Met.*, 1981, vol.24, pp.7-11.
49. W. J. Huppmann and G. Petzow, "The Elementary Mechanisms of Liquid Phase Sintering," *Sintering Processes*, G. C. Kuczynski (ed.), Plenum Press, New York, NY, 1980, pp.189-201
50. Z. S. Nikolic, M. M. Ristic, and W. J. Huppmann, "A Simple Method for Computer Simulation of Liquid Phase Sintering," *Modern Developments in Powder Metallurgy*, vol.12, H. H. Hausner, H. W. Antes, and G. D. Smith (eds.), Metal Powder Industries Federation, Princeton, NJ, 1981, pp.479-502.
51. E. G. Zukas, P. S. Z. Rogers, and R. S. Rogers, "Spheroid Growth by Coalescence During Liquid-Phase Sintering," *Z. Metallkde.*, 1976, vol.67, pp.591-595.
52. D. K. Yoon and W. J. Huppmann, "Chemically Driven Growth of Tungsten Grains During Sintering in Liquid Nickel," *Acta Met.*, 1979, vol.27, pp.973-977.
53. N. M. Parikh and M. Humenik, "Cermets: II, Wettability and Microstructure Studies in Liquid-Phase Sintering," *J. Amer. Ceramic Soc.*, 1957, vol.40, pp.315-320.
54. W. A. Kaysser, S. Takajo, and G. Petzow, "Low Energy Grain Boundaries in Liquid Phase Sintered Cu-Ag," *Z. Metallkde.*, 1982, vol.73, pp.579-580.
55. A. N. Niemi, L. E. Baxa, J. K. Lee, and T. H. Courtney, "Coalescence Phenomena in Liquid Phase Sintering - Conditions and Effects on Microstructure," *Modern Developments in Powder Metallurgy*, vol.12, H. H. Hausner, H. W. Antes, and G. D. Smith (eds.), Metal Powder Industries Federation, Princeton, NJ, 1981, pp.483-495.
56. S. Takajo, "Particle Growth by Coalescence During Liquid Phase Sintering of Fe-Cu and Cu-Ag," Ph.D. Thesis, University of Stuttgart, Stuttgart, FRG, 1981.
57. W. J. Huppmann and G. Petzow, "The Role of Grain and Phase Boundaries in Liquid Phase Sintering," *Ber. Bunsenges. Phys. Chem.*, 1978, vol.82, pp.308-312.
58. T. H. Courtney, "Microstructural Evolution During Liquid Phase Sintering: Part I. Development of Microstructure," *Metall. Trans. A*, 1977, vol.8A, pp.679-684.
59. T. H. Courtney, "Densification and Structural Development in Liquid Phase Sintering," *Metall. Trans. A*, 1984, vol.15A, pp.1065-1074.
60. R. Warren, "Particle Growth During Liquid Phase Sintering," *Inter. J. Powder Met. Powder Tech.*, 1977, vol.13, pp.249-252.
61. T. K. Kang and D. N. Yoon, "Coarsening of Tungsten Grains in Liquid Nickel-Tungsten Matrix," *Metall. Trans. A*, 1978, vol.9A, pp.433-438.
62. T. H. Courtney and J. K. Lee, "An Analysis for Estimating the Probability of Particle Coalescence in Liquid Phase Sintered Systems," *Metall. Trans. A*, 1980, vol.11A, pp.943-947.
63. W. J. Muster and H. Willerscheid, "Crystallographic Orientation Relationships in Coalescing Sintered Tungsten Spheres," *Metallog.*, 1979, vol.12, pp.287-294.

64. D. R. H. Jones, "The Migration of Grain-Boundary Grooves at the Solid-Liquid Interface," *Acta Met.*, 1978, vol.26, pp.689-694.
65. L. Kozma and W. J. Huppmann, "Experimental Method for Determining Transport Paths in Liquid Phase Sintering," *Inter. J. Powder Met. Powder Tech.*, 1979, vol.15, pp.115-119.
66. R. M. German, "The Contiguity of Liquid Phase Sintered Microstructures," *Metall. Trans. A*, 1985, vol.16A, pp.1247-1252.
67. R. Warren and M. B. Waldron, "Microstructural Development During the Liquid-Phase Sintering of Cemented Carbides. II. Carbide Grain Growth," *Powder Met.*, 1972, vol.15, pp.180-201.
68. O. J. Kwon and D. N. Yoon, "The Liquid Phase Sintering of W - Ni," *Sintering Processes*, G. C. Kuczynski (ed.), Plenum Press, New York, NY, 1980, pp.203-218.
69. Y. S. Kim, J. K. Park, and D. N. Yoon, "Liquid Flow into the Interior of W-Ni-Fe Compacts During Liquid Phase Sintering," *Inter. J. Powder Met. Powder Tech.*, 1985, vol.21, pp.29-31.
70. S. J. L. Kang, W. A. Kaysser, G. Petzow, and D. N. Yoon, "Elimination of Pores During Liquid Phase Sintering of Mo-Ni," *Powder Met.*, 1984, vol.27, pp.97-100.
71. W. Rutkowski, "Quelques Problemes du Frittage en Presence de la Phase Liquide et D'Infiltration," *Planseeber. Pulvermet.*, 1973, vol.21, pp.164-176.
72. W. A. Kaysser, and G. Petzow, "Geometry Models for the Elimination of Pores During Liquid Phase Sintering in Systems with Incomplete Wetting," *Sci. Sintering*, 1984, vol.16, pp.167-175.

CHAPTER SIX

Final Stage Processes: Microstructural Coarsening

A. Overview

As liquid phase sintering progresses, densification slows while microstructural coarsening continues. The maximum density attained in the final stage is highly dependent on the characteristics of the pores and any internal gases trapped in the pores. Futhermore, the skeletal microstructure typical to the final stage provides rigidity to the compact, and thereby inhibits pore elimination. The typical compact has already begun to exhibit considerable grain growth by the onset of the final stage. With continued sintering, the grain size further enlarges by a process of solution-reprecipitation. During grain growth there are simultaneous changes in the grain size distribution, pore size, intergranular neck size, grain-matrix interfacial area, number of solid grains, and mean grain separation.

In the final stage, the microstructure will approach a minimum energy solid-liquid configuration. The grain and liquid shape will depend on the surface energies and volume fraction of liquid. For a system with an isotropic surface energy, the solid-liquid equilibrium shape is determined by the dihedral angle as discussed in Chapter 2 (1-4). The amount of densification is inversely dependent on the dihedral angle. As a consequence a dihedral angle over 60° and a small quantity of liquid will retard final densification and give a noncontinuous liquid structure dispersed along grain edges. More practical interest exists in systems with dihedral angles below 60°. The minimum energy grain shapes will be established during the final stage as pores are eliminated and solution-reprecipitation continues to give grain growth and grain shape accommodation.

The microstructural changes observed in the final stage influence properties like wear resistance, strength, fracture toughness, electrical arcing characteristics, magnetic behavior, and ductility (5). Because microstructure is important to the properties of liquid phase sintered materials, control of the changes during the final stage is desirable. The attainment of full density is beneficial to the sintered properties. However, concomitant grain growth can degrade properties (in a manner analogous to that observed with structural metals and ceramics). This chapter discusses the various densification and coarsening phenomena evident in the final stage of liquid phase sintering.

B. Densification

During the period of rapid densification in the early stages of liquid phase sintering, the solid grains are in poor contact. By the final stage, a rigid solid skeleton has formed which inhibits rapid densification (6). In spite of the solid skeleton, solution-reprecipitation continues, giving grain growth, grain shape accommodation, and final pore removal. During the final stage, the pores can be treated as isolated spheres and the total porosity is less than 8% (7-10). A typical microstructure is shown in Figure 6.1 for a W-Ni-Fe alloy sintered at 1480°C, with spherical final pores. In the typical liquid phase sintering compact, the final stage densification rate can be estimated as follows (10):

$$d\rho/dt = 3\ D\ C\ \Omega/(k\ T\ R^2)\ [\xi/(1 + \xi)]\ (2\ \gamma/r - P_p) \tag{6.1}$$

where ρ is the fractional density, t is the time, D is the diffusivity of the solid in the liquid, C is the solid concentration in the liquid, Ω is the atomic volume, k is Boltzmann's constant, T is the absolute temperature, R is the solid grain size, ξ is a geometric term defined below, γ is the liquid-vapor surface energy, r is the pore radius, and P_p is the gas pressure in the pore.

The dimensionless geometric term ξ in Equation (6.1) depends on the grain size, pore size, and number of pores per unit volume N as follows:

$$\xi = 4\ \pi\ N\ R^2\ r/\ 3. \tag{6.2}$$

For the typical case of final stage sintering, ξ approaches zero as r, the pore radius, becomes small. Because of the high diffusivity through the liquid phase, densification in the final stage should be rapid as the pore size and porosity decrease. However, several factors can inhibit final densification (11). These include trapped gas in the pores, decomposition products from the sintering components, gross imperfections in packing, and reaction products involving the atmosphere. Sintering in a nonsoluble atmosphere will result in a trapped gas in the pores which will inhibit densification. Densification stops when the increasing pore pressure (due to densification) equals the surface tension force, that is,

$$P_p = 2\ \gamma/r. \tag{6.3}$$

This leads to a limiting final porosity. If the pores pinch closed with an insoluble ideal gas atmosphere at a pressure of P_o and a porosity of ε_o, then the minimum porosity ε_{min} is

$$\varepsilon_{min} = 0.172\ (P_o\ \varepsilon_o/\gamma)^{3/2} N^{-1/2}. \tag{6.4}$$

Equation (6.4) can also be expressed in terms of the initial pore radius when the pores close r_o

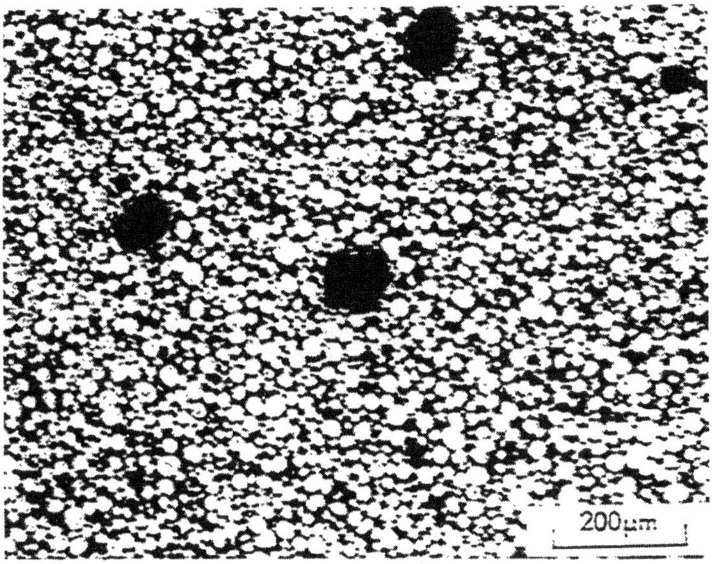

Figure 6.1 Unfilled spherical pores in a tungsten heavy alloy sintered at 1480°C (photo courtesy of E. G. Zukas).

$$\varepsilon_{min} = \varepsilon_o \, [P_o \, r_o/(2 \, \gamma)]^{3/2}. \tag{6.5}$$

The porosity at pore closure (when the gas becomes trapped in the pores) is typically about 8%. If the pore radius scales with the particle size, then Equation (6.5) predicts a lower final porosity with smaller particles. Again, this demonstrates a benefit from smaller initial powder sizes.

The consequence of a trapped gas is a final density less than the theoretical density. If the trapped gas is totally insoluble in the material, then a final density of approximately 99% of theoretical is common. The actual value of the final porosity is given by Equations (6.4) and (6.5). Alternatively, if the trapped gas is soluble in the matrix, then final densification may be impeded by the dissolution of the gas from the closed pores; but full density is possible (10). In this case, the rate of densification will be controlled by the rate of gas diffusion out of the pore or the solubility of the gas in the matrix. For the case of gas diffusion controlled pore collapse, Figure 6.2 illustrates the effect of gas diffusivity through the matrix on the density. This figure shows the calculated density in the final stage of liquid phase sintering for a typical sintering system. The faster the gas diffuses through the matrix, then the less impediment it offers to final densification. If the sintering is performed under vacuum and no decomposition products are released, then densification occurs unimpeded by trapped gas. Figure 6.3

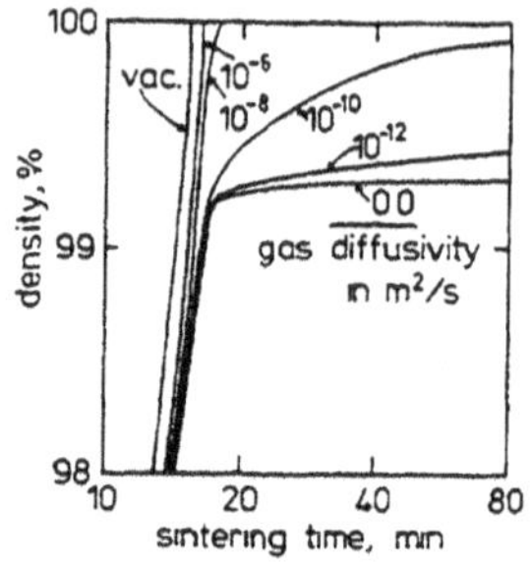

Figure 6.2 Final stage density versus sintering time for a compact with various gas diffusivities through the matrix phase (10).

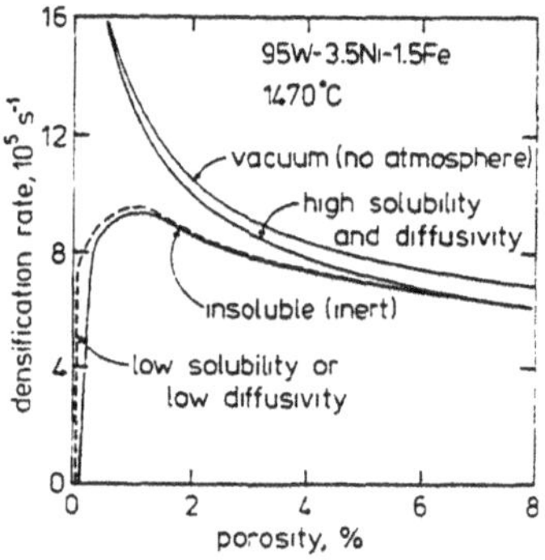

Figure 6.3 The effect of the pore atmosphere on the final stage densification rate as the porosity decreases; calculated for a tungsten heavy alloy (10).

compares four cases (vacuum, soluble gas, low solubility gas, and insoluble gas) by showing the densification rate versus the porosity for a typical liquid phase sintered material. It was assumed that the pores closed at 8% porosity and densification followed Equation (6.1). Accordingly, with no pore pressure the densification rate increases as the pore volume decreases. The soluble gas case is typical to sintering metals in a hydrogen atmosphere or oxide ceramics in an oxygen atmosphere.

Another source of gas is through decomposition products from the sintering material. The release of reaction products with a high vapor pressure will cause swelling (11,12). Figure 6.4 shows an example of this problem for silicon nitride sintered with a magnesia liquid. The sintered density is plotted versus the sintering temperature for two sintering times. The density exhibits a complex dependence on the sintering time-temperature

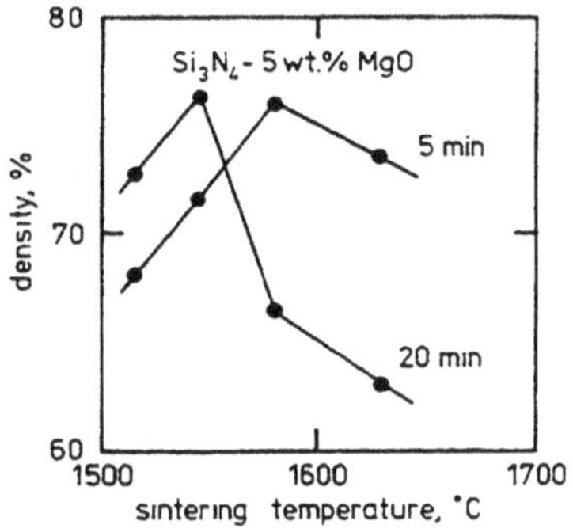

Figure 6.4 Decomposition effects on the density for liquid phase sintered silicon nitride. Higher temperatures and longer times increase the amount of decomposition induced swelling (12).

combination due to simultaneous densification and decomposition. Terwilliger and Lange (12) modeled this process assuming the decomposition vapor caused pore growth. The consequence of an increasing amount of vapor was to raise the pore pressure and thereby limit the sintered density. Swelling occurred because the decomposition reaction increased the pore size and net porosity.

In many instances of liquid phase sintering, the sintered density reaches 99% of theoretical within short sintering times as illustrated in Figure 6.5. This figure plots the porosity versus sintering time for TiC-Co sintered at 1400°C in a low pressure argon atmosphere. Near full density is attained within the first 40 minutes at the sintering temperature. The time to reach a high density obviously varies with several parameters, but prolonged sintering clearly gives a decrease in the sintered density. Likewise, Figure 6.6 shows near full density for W-Ni-Fe compacts at short sintering times in an atmosphere of pure hydrogen. In both Figures 6.5 and 6.6, the density decreases with prolonged sintering beyond one hour. The maximum density is determined by the pore size, atmosphere, surface energy, and pore density. During prolonged heating, both grain growth and pore growth are observed. Pore growth is by an Ostwald ripening mechanism because the atmosphere in the smaller pores has a greater solubility in the matrix than that of the larger pores (see Equation (3.13)). As a consequence of the differing solubilities, the large pores grow at the expense of the small pores, giving a net increase in mean pore size and a decrease in the pore pressure. Thus, as shown in Figures 6.5 and 6.6, the density decreases with prolonged sintering times. Pore growth by Ostwald ripening is analogous to solution-reprecipitation controlled grain growth discussed later in this chapter. The effect of pore coarsening is an increase in the mean pore radius and a decrease in the pore pressure as determined by Equation (6.3). Consequently, the porosity will increase with time after reaching a minimum value. Such behavior has been documented in several past studies (6,10,13-15). Microstructural analysis indicates pore coarsening occurs simultaneously with swelling during final stage sintering. Such a process requires the pores be filled with a gas which has some solubility in the matrix, and

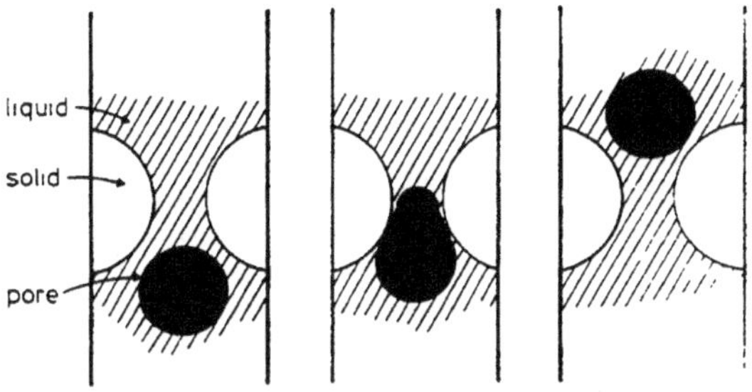

Figure 6.7 Pore migration through the compact is caused by buoyancy forces, but eventual exit is restricted by the intergranular separation.

The porosity will increase with time as a consequence of coalescence. This also results in a decreasing density and degraded mechanical properties. Eventually, the porosity will decrease due to pore exit from the compact (17). For pores to exit from the compact, they must not be much larger than the mean separation between solid grains. Thus, the densification behavior of a material sintered in a soluble gas atmosphere appears as shown in Figure 6.8. In the early portion of the final stage, the pore size is decreasing and the rate of densification is high. As the pores collapse, the trapped gas becomes pressurized and inhibits densification. Since pore coarsening takes place along with densification, it is possible for the densification process to end and swelling to take over. Eventually, the pores coarsen, detach, and rise through the compact because of buoyancy, giving densification once again. Figure 6.5 illustrates this late stage densification behavior for a compact of W-Ni-Fe sintered at 1450°C in hydrogen (14). The minimum in porosity was 0.06% at approximately 30 minutes of sintering. The density decreased during subsequent sintering until a porosity of 1.9% was obtained after 480 minutes. Finally, after 44 hours a porosity of 0.5% was obtained. This behavior agrees with that shown schematically in Figure 6.8.

Pores are typically detrimental to the properties of liquid phase sintered materials. Pore elimination in the final stage of sintering is difficult whenever the pores contain a vapor. Often densities of 99% of theoretical are the maximum attainable without special efforts to remove the residual pores. Stable pore structures can be generated by several mechanisms during prolonged sintering. Pore growth and consequential swelling are possible by Ostwald ripening and coalescence. Vacuum sintering avoids these problems. Alternatively, external pressurization can collapse the pores. This latter technique is in routine use as a last step in the liquid phase sintering of cemented carbides. It is noteworthy that superior properties are attributed to the elimination of the residual porosity present after normal liquid phase sintering.

C. Grain Growth

For most systems undergoing liquid phase sintering, coarsening of the microstructure occurs in parallel with densification. In the final stage of

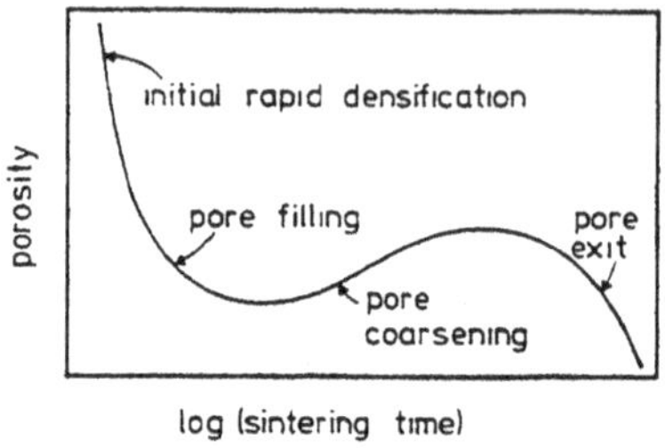

Figure 6.8 The porosity variation with sintering time. Initially pores collapse, but during prolonged sintering coarsening leads to swelling until the pores exit the compact.

sintering, the coarsening processes are dominant. The driving force for coarsening is a decrease in the interfacial energy through a reduction in both the amount and curvature of the solid-matrix interface. The chemical potential of a solid grain varies with the the radius of curvature, as discussed in Chapter 3. Grains of small dimensions have small radii of curvature and are more soluble in the liquid than grains of large dimensions. A reduction in chemical potential is possible by decreasing the curvature; that is, by enlarging the scale of the microstructure by increasing the mean radius. An easy measure of the microstructural coarsening is the grain size. With time the mean grain size increases since the large grains grow and the small grains disappear even though the volume fraction of solid remains fairly constant. This leads to a decrease in the interfacial area. Figure 6.9 shows the sintered microstructures for a 93 wt.% tungsten alloy sintered at two different temperatures for 2 hours. The magnifications are the same and the quantities of liquid phase are similar, yet the grain sizes are very different. In spite of similar volume fractions of solid in these two samples, there are differences in the number of grains per unit volume, grain-matrix surface area, and separation between grains. The two structures could be made to appear similar by using differing magnifications. For this reason, the final stage is often discussed in terms of microstructural scaling since no other effects are evident after pore removal.

The problems associated with microstructural coarsening during liquid phase sintering have been considered on several occasions (18-23). For this presentation, the primary focus will be on grain growth, with a subsequent discussion on the other microstructural factors. The kinetics of grain growth give the mean grain size versus time as follows:

$$\bar{R} \sim t^{1/n} \tag{6.6}$$

where $\bar{R}$ is the mean grain size, t is the isothermal time, and n varies from 2 to 4 for most pure materials (25). Although Equation (6.6) gives an adequate description of average grain growth, the behavior of a specific grain depends on its local environment. For liquid phase sintering, grain growth occurs by transport through the liquid. An analytic solution is available for the case

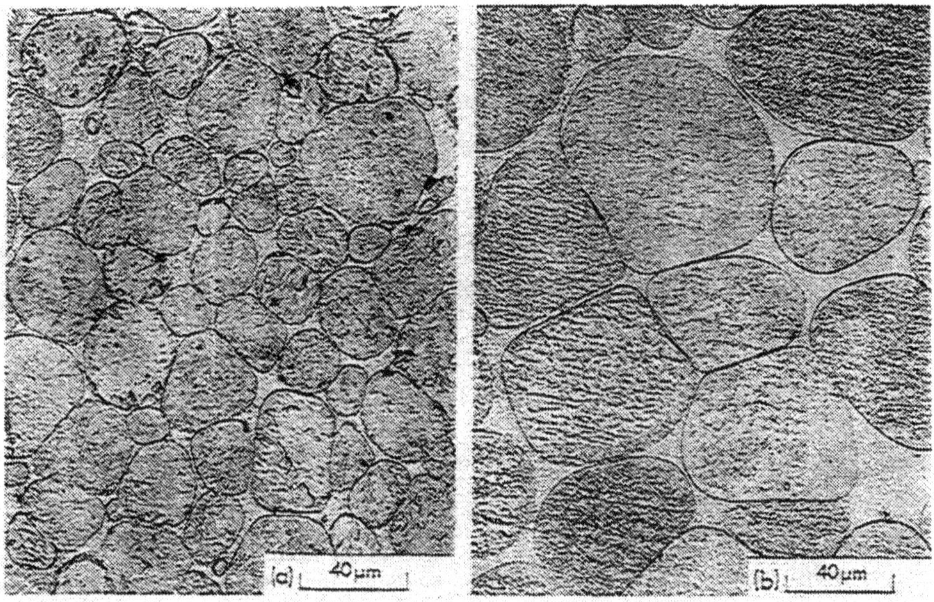

Figure 6.9 Micrographs showing grain coarsening in two 93 wt. % tungsten heavy alloys sintered at 1490°C (a) and 1580°C (b) for 2 hours (photos courtesy of L. L. Bourguignon).

where the volume fraction of solid approaches zero (infinite dilution). However, the situations typical to liquid phase sintering are poorly treated by the current models because of assumed spherical grains without contacts (zero contiguity). In general, for solution-reprecipitation controlled grain growth, these models predict the exponent n in Equation (6.6) as 3.

Several derivations of the grain coarsening kinetics have been presented in the past (23-39). Differences in these derivations will not be detailed here except as they relate to the effect of volume fraction solid. There are two approaches to the grain growth problem; i) continuum with no details on the diffusion field, and ii) discrete with consideration of neighbor-neighbor interactions. In the continuum approach for a dilute solid concentration, Greenwood (27) has summed the loss and gains over the grains. His result gives the rate of change in size for an individual spherical grain of radius R as follows:

$$dR/dt = K\, R^{-1}\, (\bar{R}^{-1} - R^{-1}) \tag{6.7}$$

where K is a rate (kinetic) constant given as follows:

$$K = 2\, D\, C\, \Omega\, \gamma/(k\, T) \tag{6.8}$$

with D equal to the diffusivity of the solid through the liquid, C equal to the concentration of the solid in the liquid, Ω equal to the atomic volume of the solid, γ equal to the solid-liquid surface energy, k equal to Boltzmann's constant, and T equal to the absolute temperature. The units of K are volume per unit time. Because both diffusivity and solubility have exponential temperature dependencies, the kinetic constant K is very sensitive to temperature.

Figure 6.10 is a schematic plot of Equation (6.7) for three different mean grain sizes. Grains with a size smaller than average will shrink, while grains larger than the average size will grow. The maximum in the growth rate occurs for a grain of size twice the average size, with a rate of growth as

$$dR/dt_{max} = K/(4\ \bar{R}^2). \tag{6.9}$$

According to Equation (6.9), the maximum growth rate declines as coarsening continues. For diffusion controlled growth with no neighbor-neighbor interactions, several assumption are made concerning the coarsening system. These assumptions include an isotropic surface energy, no contact between grains, spherical grain shape, quasi-stationary diffusion field, and a mean concentration of solid in the liquid. The classic solution for the grain size versus time under diffusion controlled growth at infinite solid dilution is given as,

$$\bar{R}^3 = \bar{R}_o^3 + 4\ K\ t/9 \tag{6.10}$$

with $\bar{R}_o$ equal to the initial mean grain size.

A prediction of the classic solution is that the maximum grain size will be 1.5 times the mean grain size under steady-state growth conditions. After coarsening has progressed, the initial mean grain size term in Equation (6.10) becomes relatively small, giving the general form shown by Equation (6.6) with n equal to 3.

It is possible that grain growth will be controlled by the interfacial reaction at either the dissolving grain or the growing grain. Such a situation arises in impure systems and systems having multiple diffusing species, where the diffusion rate is fast in comparison to the reaction rate. In these cases, the predicted grain size versus time relation will be given as follows:

$$\bar{R}^2 = \bar{R}_o^2 + 64\ \gamma\ C\ \Omega\ k_r\ t/(81\ k\ T) \tag{6.11}$$

where k_r is the interfacial reaction rate constant.

For reaction control, the activation energy for the interfacial reaction rate constant will be much larger than that for diffusion through the liquid. Thus, reaction controlled growth will be more sensitive to temperature changes than diffusion controlled growth.

Diffusion controlled growth is most typically observed in liquid phase sintering; although, there are a few reports of reaction controlled growth.

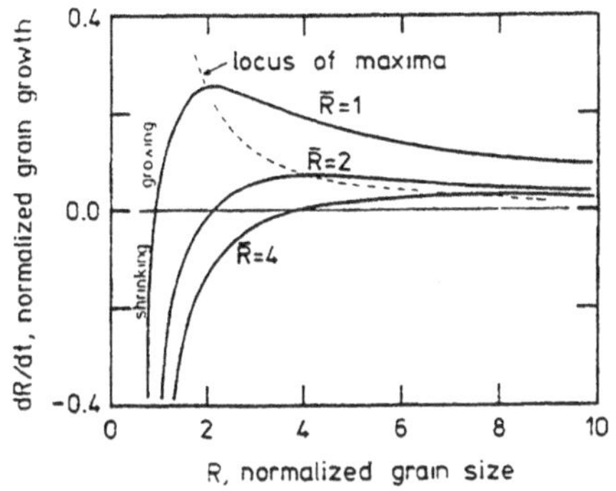

Figure 6.10 The grain growth rate versus the grain size for three different mean grain sizes, nominally 1, 2, and 4.

Table 6.1 tabulates the results for several systems involving grain growth observations (10,18,33,40-57). Reaction control is typically observed in the more complex systems involving several components. These are usually based on limited analysis of growth characteristics. However, a clear conclusion as to the controlling mechanism is often not easy, since actual systems differ considerably from the simple models. In several of the past analyses, the conclusion with respect to the controlling mechanism has been based on grain size distribution characteristics. As will be demonstrated in the next section, there is considerable uncertainty as to the grain size distribution; thus, this is not a valid basis for identification of the coarsening mechanism. Hence, some of the results shown in Table 6.1 are of questionable validity.

The effect of a grain size distribution is for the small grains to shrink while the large grains grow. The grain shape often is not spherical, but a form similar to Equation (6.10) is obeyed since the volumetric growth rate is the dominant feature. Voorhees and Glicksman (21,23,37) have computer simulated the coarsening of a multiple particle system. They determined the behavior of individual grains versus time, volume fraction solid, and grain size distribution. The behavior shows the expected shrinkage of small grains with the simultaneous growth of large grains. The grain growth exponent has been found to be 3 for all cases. Although there is a neighbor effect on the growth or shrinkage of an individual grain, the overall system behavior appears to converge to an environment independent form.

The effect of a high solid volume fraction is to accelerate the rate of grain growth because the diffusion distance decreases. An analysis of grain growth by diffusion control versus the volume fraction of solid phase has been the subject of several theories (24,26-39). Various assumptions are necessary concerning the diffusion field, precipitate redistribution, concentration gradients, coalescence, initial size distribution, and precipitate shape. The basic result is a kinetic law similar to Equation (6.10) with cube of the grain size varying with time. The rate constant K is modified to account for the effect of the shorter diffusion distance as the volume fraction of solid

TABLE 6.1

Grain Growth Mechanisms

During Liquid Phase Sintering
(references in parentheses)

Diffusion Control		Reaction Control	
Mo-Ni-Fe	(40)	PbS-NaCl-KCl	(52)
$MgO-V_2O_5$	(41)	$SmCo_5$-Sm-Co	(53)
Fe-Cu	(18, 42, 43)	TiC-Mo-Ni	(54)
Co-Cu	(43, 44)	NbC-Fe-B	(33)
Cu-Ag	(18)	WC-Co	(55)
W-Ni-Cu	(57)	TiN-TiC-Ni	(56)
W-Ni-Fe	(10, 18, 45, 46)		
W-Ni	(47)		
TiC-Co	(48)		
HfC-Co	(48)		
TaC-Co	(48)		
Mo_2C-Co	(48)		
NbC-Co	(33)		
CaO-MgO-Fe_2O_3	(50)		
Pb-Sn	(51)		
VC-Co	(48, 49)		

increases. Several forms of this modification have been suggested (39). Figure 6.11 compares the predicted dependencies of the growth rate constant on the volume fraction of solid from several studies. All of the results suggest the rate of grain growth will increase as the volume fraction of solid increases, in agreement with experimental results.

Figure 6.12 demonstrates the effect of an increasing volume fraction of solid on the grain size for tungsten grains during sintering in a nickel-rich liquid at 1540 °C (47). As the amount of tungsten increases, the volume fraction of solid is increased and the grain size for a given time is larger. For all three tungsten contents, the growth rate follows a time to the 1/3 power dependence. Similar results have been demonstrated for several other systems; the grain growth rate increases as the amount of solid increases.

Figures 6.13 through 6.18 further demonstrate the processing effects on grain growth for various systems. In Figure 6.13, the mean tungsten grain size is shown versus the tungsten content for a W-Ni-Fe alloy sintered at

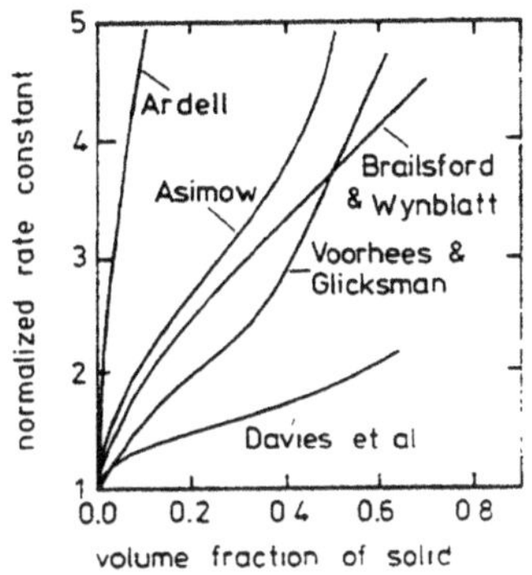

Figure 6.11 The normalized rate constant versus the volume fraction of solid according to several theories (34).

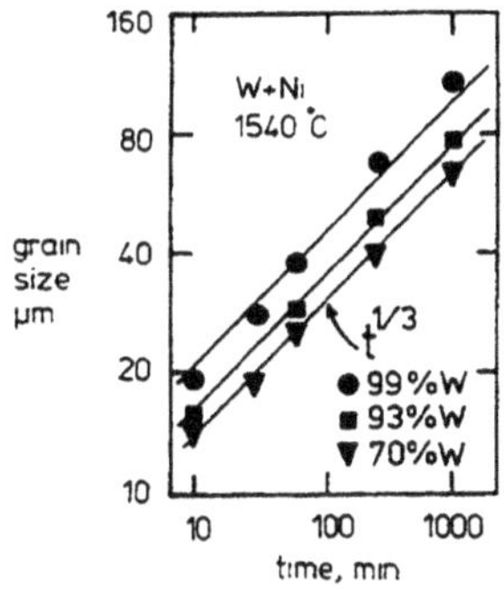

Figure 6.12 The grain size versus sintering time for tungsten sintered in a nickel-rich liquid at 1540°C, showing cubic growth for three tungsten contents (47).

1460°C for 1 hour (46). These data exhibit a strong volume fraction solid effect on grain growth. Figure 6.14 plots the cube of the mean grain size versus time for NbC-Co compacts sintered at four temperatures (48). The coarsening follows the behavior expected for final stage liquid phase sintering with faster grain growth at higher temperatures. Figure 6.15 shows the grain growth rate constant versus the inverse absolute temperature for Co-Cu alloys (43,44). The prediction of faster grain growth with a higher solid content is evident in this figure. An exponential temperature dependence exists for the rate constant because diffusion is thermally activated. If the temperature effect on solubility is known, then an activation energy can be calculated using data such as shown in Figure 6.15. For several cases, the calculated activation energies are reasonable for diffusion through the liquid,

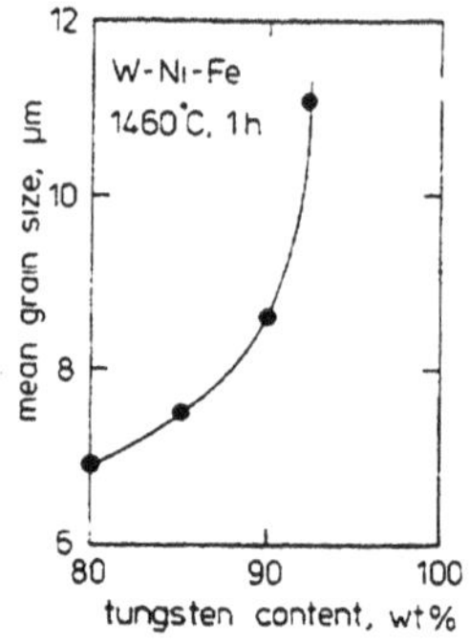

Figure 6.13 The effect of an increasing volume fraction solid on the mean grain size for W-Ni-Fe alloys given a constant processing cycle (46).

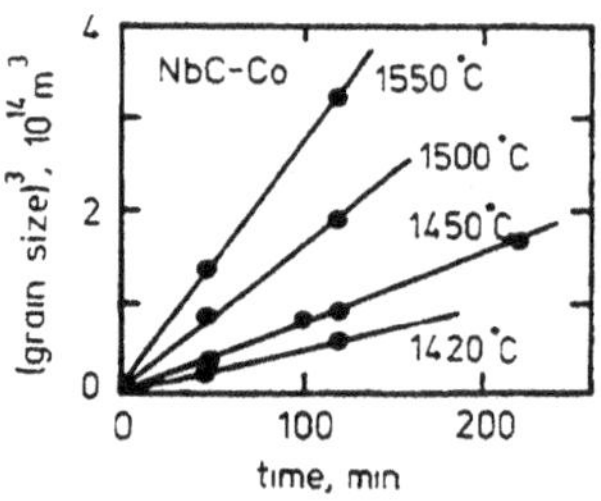

Figure 6.14 The cube of the mean grain size versus sintering time for NbC-Co compositions at four temperatures. An increase in temperature raises the rate of grain growth (48).

supporting the fundamental coarsening concept. A critical analysis of the various models is possible by comparing the predicted variations in the rate constant with the volume fraction of solid. As shown in Figure 6.16 for VC-Co compacts, a large quantity of liquid decreases the rate constant. These results suggest an inverse relation between the rate constant and the volume fraction of liquid. Figure 6.17 plots several measurements for Fe-Cu alloys versus the form predicted by Voorhees and Glicksman (34). Values from several studies were used with the rate constant at 1150°C being selected for this comparison. The measured rates were fit using their model and one adjustable parameter. This reasonable agreement between the model and experiment supports the use of their model in estimating grain growth rates during liquid phase sintering at high volume fractions of liquid.

In multiple component solid systems (such as cemented carbides), the grain growth rate will be controlled by diffusion of the slower solid species. To preserve stoichiometry in the reprecipitated material, both species must

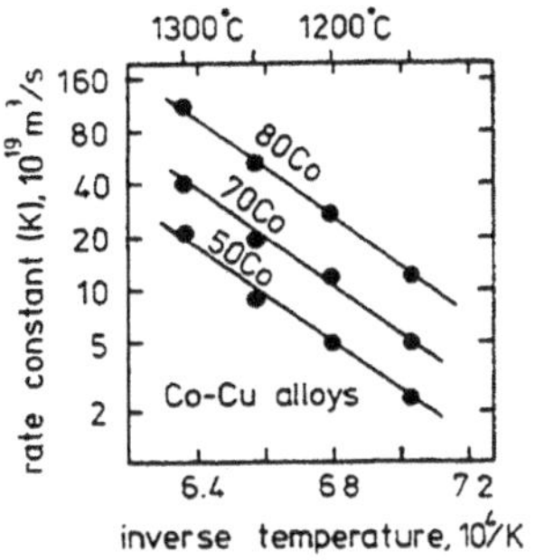

Figure 6.15 The grain growth rate constant shown as a function of the inverse temperature for Co-Cu alloys of three different cobalt contents (43).

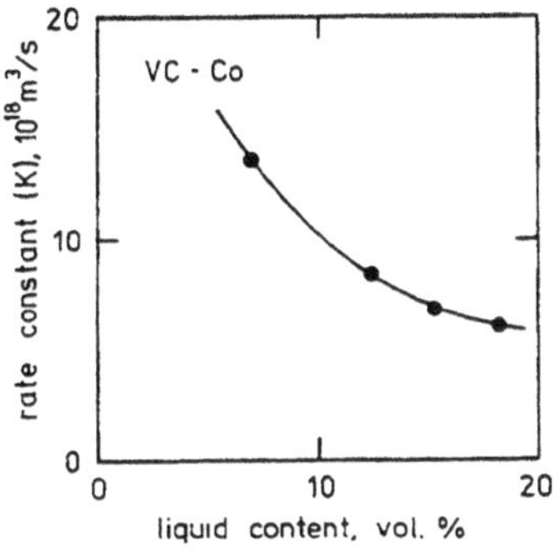

Figure 6.16 The grain growth rate constant decreases as the volume of liquid increases for vanadium-carbide sintered in a cobalt-rich liquid (49).

have coupled fluxes (22). The effective diffusion coefficient depends on the abundance of each component and the relative diffusivity. It is predicted that situations can arise where an impurity can control the grain growth rate; especially if the impurity is soluble in the solid and has a slow diffusion rate (55).

Many liquid phase sintering systems have a nonzero dihedral angle. As a consequence, necks will form between the grains. The contiguity will increase as both the volume fraction of solid and dihedral angle are raised. The effect of contiguity is to reduce the grain surface area in contact with the liquid and to slow the rate of grain growth (40,48,50,58). Such an effect has been noted in the studies by Buist et al. (58) on liquid phase sintered magnesia. Figure 6.18 plots the grain growth rate constant versus the dihedral angle as measured at two temperatures. An increase in the dihedral

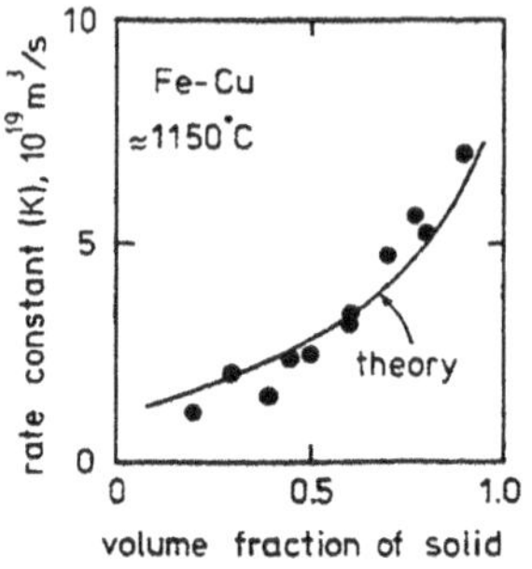

Figure 6.17 The volume fraction solid effect on the grain growth rate constant for Fe-Cu alloys as compared with the theory of Voorhees and Glicksman (34).

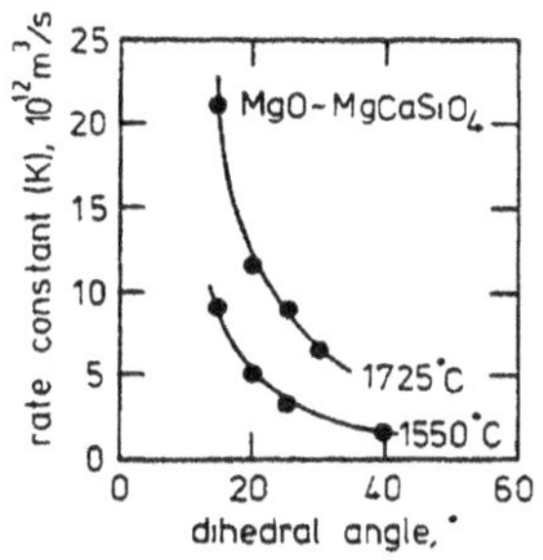

Figure 6.18 The grain growth rate constant decreases as the dihedral angle (and contiguity) increase for magnesia sintered in a $MgCaSiO_4$ liquid (58).

angle gives a higher contiguity and slows the rate of grain growth. This effect is also demonstrated by the plot in Figure 5.18, where the grain growth rate is given versus the contiguity for several carbide systems. Besides slowing grain growth, a high contiguity contributes to a rigid solid skeleton, thereby inhibiting final densification. Warren (49) suggests that the kinetic coefficient K in Equation (6.8) should be modified to include the contiguity. However, this modification has not been tested experimentally.

Coalescence, as treated in Chapter 5, has often been proposed as a mechanism of grain growth. Coalescence will broaden the grain size distribution and raise the rate constant for grain growth (18,59,60). Generally, the coalescence contribution to coarsening should increase as the volume fraction of solid and dihedral angle increase. The exponent n in Equation (6.6) will remain as 3 even with a substantial coalescence contribution. However, as noted in the last chapter, coalescence appears not to be a significant coarsening mechanism in the final stage of liquid phase sintering.

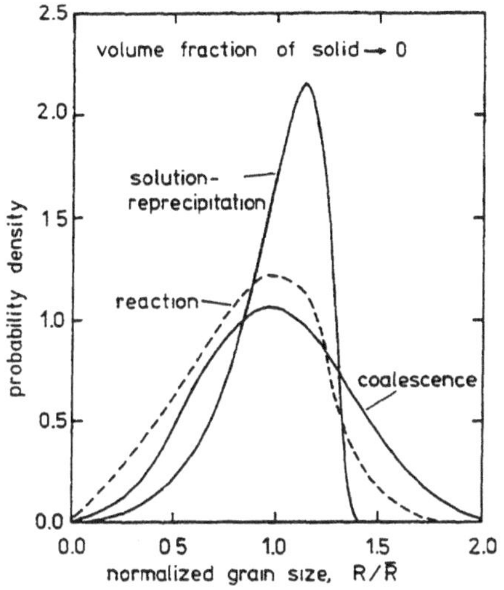

Figure 6.19 The predicted grain size distributions at zero volume fraction of solid for three potential mechanisms of grain growth.

D. Grain Size Distribution

The grain size distribution is also a subject of concern during the final stage of liquid phase sintering. The grain growth models predict various grain size distributions during steady-state grain growth. For the limiting case of zero volume fraction solid, the predicted grain size distributions are shown in Figure 6.19. There are three potential cases; growth by diffusion control, reaction control, or coalescence (24,26,29,30,49,60,61). The coalescence distribution is the broadest. All three cases correspond to spherical grains with isotropic growth patterns and infinitely dilute solutions.

As the volume fraction of solid increases, the predictions for the grain size distribution are less exact. It is generally agreed that the grain size distribution becomes broader as the volume fraction of solid increases. Also deviations from a spherical grain shape or a small contribution from coalescence will lead to a broader size distribution (43,44,60). The models predict the maximum grain size will be approximately 2.5 to 2.7 times the mean grain size for high volume fractions of solid. Observations on several carbides show the maximum grain size typically averages 2.13 times the mean grain size after prolonged sintering (48). The data of Krock (46) for W-Ni-Fe alloys are plotted in Figure 6.20. He measured the ratio of the maximum to mean grain sizes during grain coarsening at 1460°C. As shown, the ratio of maximum to mean size remained fairly constant during the time from 10 minutes to 8 hours (average value was 2.37). During this period, the mean grain size increased from 4 µm to over 13 µm, and followed the expected cubic growth law. These ratios of maximum to mean grain sizes are near the range predicted by most of the theories.

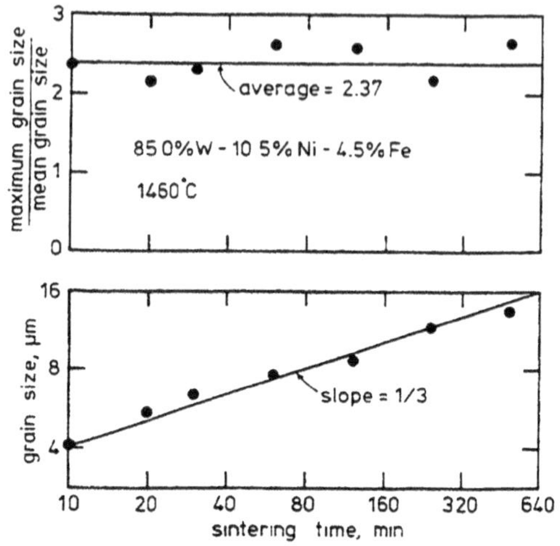

Figure 6.20 The upper plot shows the ratio of the maximum to mean grain sizes versus sintering time for a tungsten heavy alloy. The lower plot shows the concurrent grain growth (46).

Voorhees (21) has computer simulated grain growth during liquid phase sintering to calculate the grain size distribution for various volume fractions of solid, assuming isolated spherical grains. Figure 6.21 shows some results of his calculations, indicating the expected distribution broadening with increasing volume fraction of solid. Note also the maximum grain size increases as the volume fraction of solid increases.

Several attempts have been made to compare experimental distributions with the predictions using metallographic measured grain sizes. These comparisons are often the basis for conclusions concerning the controlling coarsening mechanism. Examples of such data are shown in Figure 6.22 for 95% W-3.5% Ni-1.5% Fe sintered at 1600°C for times of 1.3, 32.5 and 120 minutes (62). Unfortunately, the use of metallographic grain size data from polished cross-sections is not an appropriate basis for testing the models (22). First, the probability of sectioning a large grain is higher than that for sectioning a small grain. Second, the section is typically not through the middle of the grain, thus the apparent size is not the true size (63). Third, the grain shape often does not agree with the spherical model. Finally, most materials processed by liquid phase sintering exhibit contiguity and grain shape accommodation, which are not accounted for by the models. As a consequence, several corrections are needed in treating the data (22,55,64). In light of such corrections and the normal error in performing grain size distribution analysis, conclusions as to the coarsening mechanism are difficult to accept when based on the size distribution. At this time, the specific volume fraction effects on the grain size distribution are not clear, and data with sufficient accuracy to separate models are not available.

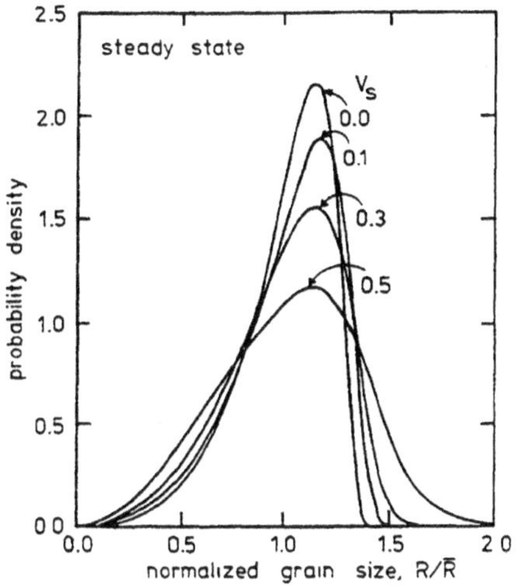

Figure 6.21 Computer simulated grain size distributions at four volume fractions of solid, showing broadening as the solid content increases (21).

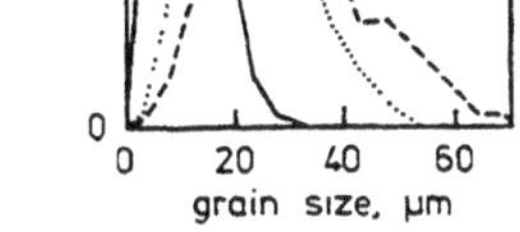

Figure 6.22 Examples of the grain size distributions after sintering for three times for W-Ni-Fe sintered at 1600°C (62).

E. Discontinuous Grain Growth

It is often observed in cemented carbides that the grain size has a nonuniform distribution (65). Coarse carbides act as seeds for rapid grain coarsening for a small fraction of the grains. This leads to a nonuniform grain size distribution and the phenomenon known as discontinuous or exaggerated grain growth.

There are two stages to discontinuous grain growth (66); nucleation of abnormal grains, and preferred growth of these grains. The properties of liquid phase sintered materials are sensitive to discontinuous grain growth. Studies involving doped systems have demonstrated discontinuous grain growth is attributable to inhomogeneities, either chemical or geometrical. Large particles in the initial powder mix act as seeds for rapid grain growth, especially at higher sintering temperatures. Additionally, agglomerates of solid particles undergo more rapid grain growth than the balance of the material. This would be expected from the dependence of the grain growth rate on volume fraction of solid discussed in the preceding section. These inhomogeneities nucleates a large grain which eventually grows into the neighboring matrix which has a lower volume fraction of solid. Impurities and off-stoichiometric grains also have been noted to cause rapid localized grain coarsening (55,65,66).

Eun (66) has considered several factors causing discontinuous grain growth in WC-Co powder mixtures. He found the evolution of the grain size distribution appeared like that shown in Figure 6.23. The normal grain coarsening process maintained the same grain size distribution shape with sintering time, with a continuous increase in the mean grain size. Alternatively, discontinuous grain growth often could be traced to a small cluster of large grains in the initial powder compact. These grains coarsened preferentially, leading to a dramatic shift in the grain size distribution.

Additionally, discontinuous grain growth has been attributed to several impurities. In the carbide systems, oxygen, carbon, boron, aluminum, phosphorus, and silicon have been implicated as causes of rapid grain growth (65-68). These impurities appear to be most detrimental when then occur inhomogeneously in the powder mix. Thus, compositional differences in the compact are potential causes of discontinuous grain growth.

F. Inhibited Grain Growth

For several materials processed by liquid phase sintering, grain growth is detrimental to the properties. As a consequence techniques to inhibit grain growth during the final stage of sintering have been developed (56,68-70). A decrease in solid solubility or diffusivity is one approach to slowing the grain growth rate. For multiple component solids (like carbides or oxides), the slower moving species will control the grain growth rate (71). A change in the stoichiometry of the solid component can result in slower grain growth if the deficient species has the lower diffusion rate. Also chemical additives can alter the interfacial energy or interfere with the interfacial dissolution and reprecipitation steps. The typical inhibitor acts to slow the interfacial material transport during solution-reprecipitation (68). As an example, cubic carbides like vanadium carbide are added to WC-Co mixtures at levels as low as 200 parts per million to inhibit grain growth of the tungsten carbide. It is unclear why a difference in crystal structure between the additive and the primary carbide results in inhibited growth. The additives show the greatest effect at the lower sintering temperatures and smaller initial grain sizes. As would be expected, the decrease in grain growth rate is accompanied by a lower densification rate in the intermediate stage.

In multiple component solid systems, inhibited grain growth is a typical observation (72,73). As two immiscible solid phases are mixed, grain growth

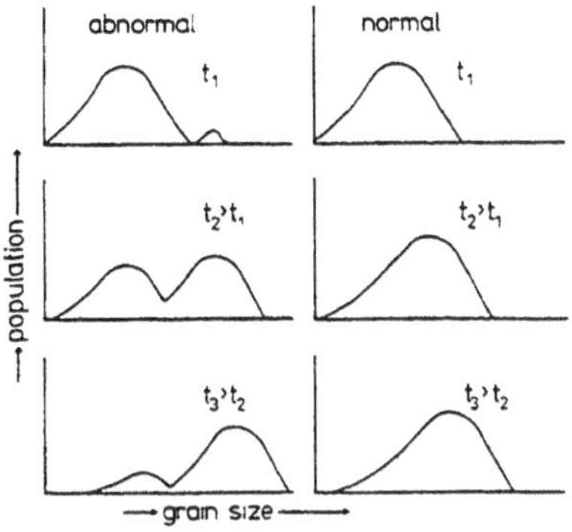

Figure 6.23 A contrast between grain size distributions during normal and abnormal grain growth; abnormal growth is initiated by the small population of coarse grains (66).

decreases (15,50,51,71-75). This effect is demonstrated in Figure 6.24 using the grain size data of White (74). The plot shows the grain size of MgO and CaO grains versus the ratio of these two solids in a liquid composed of oxides of calcium, magnesium and iron. The grain size of each component is the largest when they are present as the only solid. The presence of the second solid phase decreases the interfacial contact area for a constant volume fraction of liquid. Grain growth around the second solid phase inhibits the coarsening kinetics. The finer the size of the second solid phase, the slower the primary phase undergoes growth. This effect from mixed solid phases is probably due to the reduction in volume fractions of the respective phases, increased diffusion distances, changes in the interfacial energies, and interfacial reactions.

G. Other Microstructural Changes

As the grain size enlarges during final stage liquid phase sintering, several other microstructural changes can also be observed. The grain density is defined as the number of grains per unit volume. If the volume fraction of solid remains constant, then the number of grains per unit volume must decrease as the mean grain size increases. Assuming zero porosity and volume conservation, the cubic grain growth law leads to the conclusion that N_V, the grain density will vary as

$$N_V \sim t^{-1} \tag{6.12}$$

where t is the isothermal sintering time. Analysis of the grain population versus sintering time has substantiated this inverse time dependence (18,26,33,42). Figure 6.25 shows the data for the grain population for NbC-Fe at various temperatures (33). The faster grain growth rate at the higher temperatures gives a lower grain population. Obviously, the finer the initial particle size, the higher the grain size population for a given set of

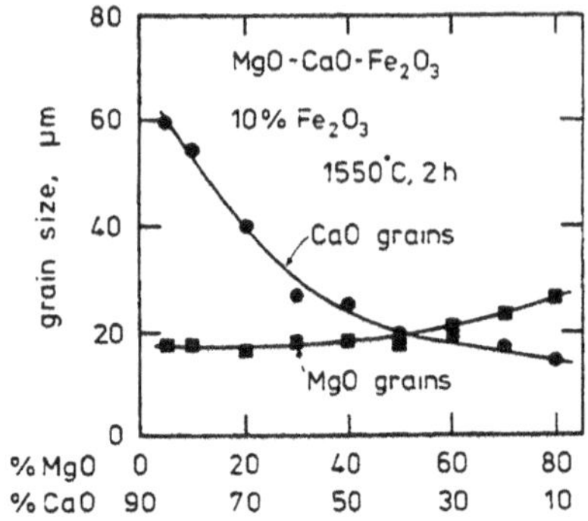

Figure 6.24 A demonstration of inhibited grain growth due to the presence of two solid phases during liquid phase sintering of $MgO\text{-}CaO\text{-}Fe_2O_3$ (74).

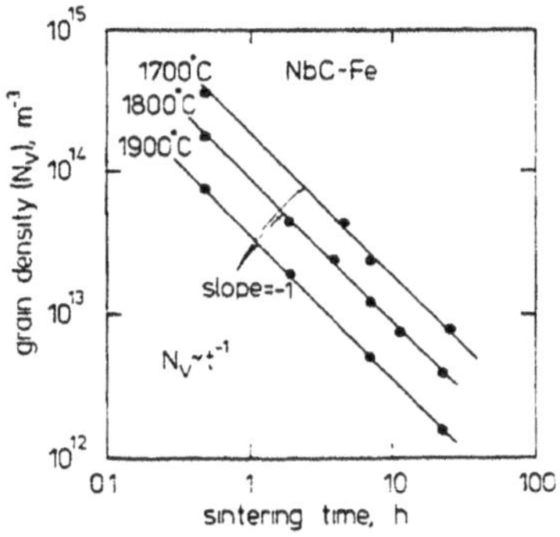

Figure 6.25 The decreasing grain density for NbC-Fe compacts sintered at three different temperatures (33).

processing conditions. However, with prolonged sintering the effect of the initial grain size is obliterated.

The grain-matrix surface area per unit volume depends on the grain size and the grain population. Accordingly, the surface area per unit volume S_V will vary as

$$S_V = 4\pi \bar{R}^2 N_V. \tag{6.13}$$

Since the grain population decreases with inverse time and the mean grain size increases with time to the 1/3 power, the interfacial surface area will decrease approximately with time to the -1/3 power. Courtney (76) has provided a more rigorous analysis, considering the volume fraction of solid V_S and the initial grain size, giving

$$S_V = S_{VO} \, [1 + K \, t/(N_C \, \bar{R}_O^3)]^{-1/3} (1 - 0.4851 \, V_S) \qquad (6.14)$$

where K is given by Equation (6.8), the packing coordination is given by N_C, and the initial surface area is S_{VO}.

Figure 6.26 shows the surface area of Fe-Cu alloys versus sintering time at three sintering temperatures (77). The decline in surface area follows the predicted -1/3 power. The higher temperature gives faster diffusion and greater reduction in surface area for a given time. During these experiments the grain size increased with time to the 1/3 power.

Another microstructural parameter of concern is the mean separation between grains. This separation depends on the grain size and volume fraction of liquid as expressed by Equation (2.13). Consequently, the grain coarsening observed in the final stage of liquid phase sintering will give a parallel increase in the mean grain separation. Again, a time to the 1/3 power is anticipated for the mean grain separation as illustrated in Figure 6.27. This figure gives the mean grain separation versus time for NbC-Fe sintered at 1800°C for four different volume fractions of solid. The mean grain separation follows the expected dependence on the cube-root of sintering time. The effect of an increase in the volume fraction of solid is to decrease the mean grain separation.

By the final stage, neck growth has established a constant neck size ratio dependent on the dihedral angle (see Equation (5.8)). Since the grain size coarsens with time to the 1/3 power, then to preserve an equilibrium dihedral angle between contacting grains, the neck size also enlarges. The observed neck size to grain size ratio remains constant; thus, the absolute neck size grows at the same rate as the grain size.

During the final stage, grain growth is very evident. In parallel with grain growth, there are concomitant changes in the grain-matrix surface area, number of grains per unit volume, mean separation between grains, and intergranular neck size. The interdependence of these parameters on one another allows predictions of the microstructural changes during liquid phase sintering. Experimentally, these parameters follow the behavior predicted for diffusion controlled grain coarsening.

H. Summary

The final stage of liquid phase sintering is an extension of the solution-reprecipitation process active during the intermediate stage. In the final stage, the structure approaches equilibrium. The microstructure is dictated by the volume fraction of solid and interfacial energies. Depending on the atmosphere, the porosity may stabilize or can be totally eliminated during the final stage. The densification rate depends on diffusion and solubility of the solid in the liquid. If there is no trapped atmosphere in the pores, then the time to full density is relatively short; under an hour. Alternatively, if an insoluble gas is trapped in the pores, then the final density is limited by the

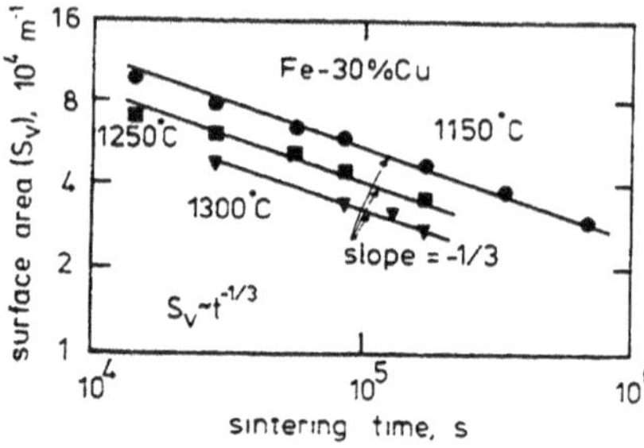

Figure 6.26 Surface area reduction versus sintering time for Fe-Cu compacts sintered at three different temperatures (77).

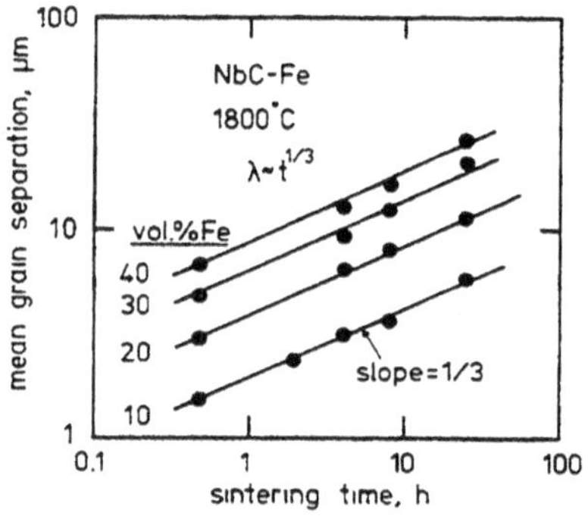

Figure 6.27 The mean grain separation versus the sintering time for NbC-Fe compacts with four iron concentrations, showing continual coarsening (33).

pore pressure. In such cases, the final density is typically 99% of theoretical. During prolonged heating, residual pores will coarsen by Ostwald ripening and coalescence mechanisms. These events can give swelling rather than densification. At the same time, the low density of the pores can lead to buoyancy driven pore migration and eventual full densification if there is sufficient liquid. Figure 6.28 sketches the density variations possible during sintering as influenced by the sintering atmosphere. For sintering in a vacuum, rapid densification is anticipated. With a soluble gas in the pores, densification is retarded by the pressurization of the trapped gas during densification. Furthermore, coarsening of the gas filled pores will give swelling at long sintering times. Finally, an insoluble gas will also impede full densification and give trapped pores filled with gas.

One of the characteristic events in the final stage of liquid phase sintering is grain growth. During grain growth, the smaller grains

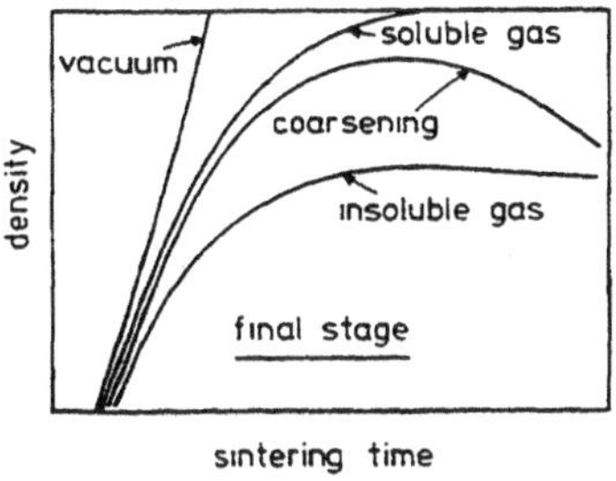

Figure 6.28 Four possible density versus time paths for final stage liquid phase sintering, where the path depends on the atmosphere characteristics.

preferentially dissolve and reprecipitate solid material on the large grains. As a consequence, the mean grain size increases with time, giving a net decrease in the number of grains and solid-liquid interfacial area. Additionally, the mean separation between the grains increases with sintering time. The kinetics of grain growth depend on the grain separation, which in turn is dependent on the grain size and volume fraction of liquid. Rapid rates of grain growth occur with small initial grain sizes, high sintering temperatures, low dihedral angles, high volume fractions of solid, and high solubilities of solid in the liquid. Diffusion through the liquid is the usual mechanism of grain growth. As a consequence the mean grain size increases with time to the 1/3 power. In complex systems, impurities or second solid phases can inhibit grain growth by controlling the interfacial reaction rate. Alternatively, discontinuous grain growth is observed in systems with inhomogeneities in chemistry, packing, or initial particle size.

The grain size distribution takes on a characteristic shape during steady-state grain growth. The distribution broadens as the volume fraction of solid increases. Many models have been proposed for both the kinetics of grain growth and for the grain size distribution. Several problems exist in comparing the models to practical liquid phase sintering data. The models do not include the effects of irregular grain shapes, high volume fractions of solid, grain shape accommodation, grain contact, and nonzero dihedral angles. Additionally, grain size distribution data can not be obtained with sufficient accuracy to critically test the models. However, the main processing effects are understood with respect to the events and microstructure development during the final stage. Thus, the effects of time, temperature, solubility, grain size, atmosphere, and composition are understood in terms of their effects on microstructural coarsening and final densification as discussed in this chapter.

I. References

1. W. Beere, "A Unifying Theory of the Stability of Penetrating Liquid Phases and Sintering Pores," *Acta Met.*, 1975, vol.23, pp.131-138.
2. P. J. Wray, "The Geometry of Two-Phase Aggregates in which the Shape of the Second Phase is Determined by its Dihedral Angle," *Acta Met.*,

1976, vol.24, pp.125-135.
3. R. Raj, "Morphology and Stability of the Glass Phase in Glass-Ceramic Systems," *J. Amer. Ceramic Soc.*, 1981, vol.64, pp.245-248.
4. Park and D. N. Yoon, "The Effect of Dihedral Angle on the Morphology of Grains in a Matrix Phase", *Metall. Trans. A*, 1985, vol.16, pp.923-928.
5. F. V. Lenel, *Powder Metallurgy Principles and Applications*, Metal Powder Industries Federation, Princeton, NJ, 1980.
6. H. S. Cannon and F. V. Lenel, "Some Observations on the Mechanism of Liquid Phase Sintering," *Plansee Proceedings*, F. Benesovsky (ed.), Metallwerk Plansee, Reutte, Austria, 1953, pp.106-121.
7. A. J. Markworth, "On the Volume Diffusion Controlled Final Stage Densification of a Porous Solid," *Scripta Met.*, 1972, vol.6, pp.957-960.
8. W. D. Kingery, "Densification During Sintering in the Presence of a Liquid Phase. 1. Theory," *J. Appl. Phys.*, 1959, vol.30, pp.301-306.
9. R. L. Coble, "Sintering Crystalline Solids. 1. Intermediate and Final State Diffusion Models," *J. Appl. Phys.*, 1961, vol.32, pp.787-792.
10. R. M. German and K. S. Churn, "Sintering Atmosphere Effects on the Ductility of W-Ni-Fe Heavy Metals," *Metall. Trans. A*, 1984, vol.15A, pp.747-754.
11. W. D. Kingery, E. Niki, and M. D. Narasimhan, "Sintering of Oxide and Carbide-Metal Compositions in Presence of a Liquid Phase," *J. Amer. Ceramic Soc.*, 1961, vol.44, pp.29-35.
12. G. R. Terwilliger and F. F. Lange, "Pressureless Sintering of Si_3N_4," *J. Mater. Sci.*, 1975, vol.10, pp.1169-1174.
13. H. Danninger, W. Pisan, G. Jangg, and B. Lux, "Veranderungen in Eigenschaften und Bruchverhalten von Wolfram-Schwermetallen Wahrend der Flussinphasensinterung," *Z. Metallkde.*, 1983, vol.74, pp.151-155.
14. K. S. Churn and D. N. Yoon, "Pore Formation and its Effect on Mechanical Properties in W-Ni-Fe Heavy Alloy," *Powder Met.*, 1979, vol.22, pp.175-178.
15. B. Meredith and D. R. Milner, "The Liquid-Phase Sintering of Titanium Carbide," *Powder Met.*, 1976, vol.19, pp.162-170.
16. E. M. Baroody, "Calculations on the Collisional Coalescence of Gas Bubbles in Solids," *J. Appl. Phys.*, 1967, vol.38, pp.4893-4903.
17. T. H. Courtney, "Densification and Structural Development in Liquid Phase Sintering," *Metall. Trans. A*, 1984, vol.15A, pp.1065-1074.
18. Y. Masuda and R. Watanabe, "Ostwald Ripening Processes in the Sintering of Metal Powders," *Sintering Processes*, G. C. Kuczynski (ed.), Plenum Press, New York, NY, 1980, pp.3-21.
19. F. F. Lange and B. I. Davis, "Sinterability of Zirconia and Alumina Powders: The Role of Pore Coordination Number Distribution," Report No. SC5325.2FR, Rockwell Science Center, Thousand Oaks, CA, May 1983.
20. H. E. Exner, "Ostwald-Reifung von Ubergangsmetallkarbiden in Flussingem Nickel und Kobalt," *Z. Metallkde.*, 1973, vol.64, pp.273-279.
21. P. W. Voorhees, "Ostwald Ripening in Two-Phase Mixtures," Ph.D. Thesis, Rensselaer Polytechnic Institute, Troy, NY, 1982.
22. H. Fischmeister and G. Grimvall, "Ostwald Ripening - A Survey," *Sintering and Related Phenomena*, G. C. Kuczynski (ed.), Plenum Press, New York, NY, 1973, pp.119-149.
23. P. W. Voorhees and M. E. Glicksman, "Solution to the Multi-Particle Diffusion Problem with Applications to Ostwald Ripening- I. Theory," *Acta Met.*, 1984, vol.32, pp.2001-2011.

24. K. Tsumuraya, "Coarsening Models Incorporating Both Diffusion Geometry and Volume Fraction of Particles," *Acta Met.*, 1983, vol.31, pp.437-452.
25. M. P. Anderson, D. J. Srolovitz, G. S. Grest, and P. S. Sahni, "Computer Simulation of Grain Growth - I. Kinetics," *Acta Met.*, 1984, vol.32, pp.783-791.
26. I. M. Lifshitz and V. V. Slyozov, "The Kinetics of Precipitation from Supersaturates Solid Solutions," *J. Phys. Chem. Solids*, 1961, vol.19, pp.35-50.
27. G. W. Greenwood, "The Growth of Dispersed Precipitates in Solutions," *Acta Met.*, 1956, vol.4, pp.243-248.
28. C. Wagner, "Theory of Precipitate Change by Redissolution," *Z. Electrochem.*, 1961, vol.65, pp.581-591.
29. A. J. Ardell, "The Effect of Volume Fraction on Particle Coarsening: Theoretical Considerations," *Acta Met.*, 1972, vol.20, pp.61-71.
30. C. K. L. Davies, P. Nash, and R. N. Stevens, "The Effect of Volume Fraction of Precipitate on Ostwald Ripening," *Acta Met.*, 1980, vol.28, pp.179-189.
31. A. D. Brailsford and P. Wynblatt, "The Dependence of Ostwald Ripening Kinetics on Particle Volume Fraction," *Acta Met.*, 1979, vol.27, pp.489-497.
32. K. W. Lay, "Grain Growth in Urania - Alumina in the Presence of a Liquid Phase," *J. Amer. Ceramic Soc.*, 1968, vol.51, pp.373-376.
33. S. Sarian and H. W. Weart, "Kinetics of Coarsening of Spherical Particles in a Liquid Matrix," *J. Appl. Phys.*, 1966, vol.37, pp.1675-1681.
34. P. W. Voorhees and M. E. Glicksman, "Ostwald Ripening During Liquid Phase Sintering- Effect of Volume Fraction on Coarsening Kinetics," *Metall. Trans. A*, 1984, vol.15A, pp.1081-1088.
35. J. J. Weins and J. W. Cahn, "The Effect of Size and Distribution of Second Phase Particles and Voids on Sintering," *Sintering and Related Phenomena*, G. C. Kuczynski (ed.), Plenum Press, New York, NY, 1973, pp.151-163.
36. R. Asimow, "Clustering Kinetics in Binary Alloys," *Acta Met.*, 1963, vol.11, pp.72-73.
37. P. W. Voorhees and M. E. Glicksman, "Solution to the Multi-Particle Diffusion Problem with Applications to Ostwald Ripening - II. Computer Simulations," *Acta Met.*, 1984, vol.32, pp.2013-2030.
38. J. A. Marquesee and J. Ross, "Theory of Ostwald Ripening: Competitive Growth and its Dependence on Volume Fraction," *J. Chem. Phys.*, 1984, vol.80, pp.536-543.
39. P. W. Voorhees, "Theory of Ostwald Ripening," *J. Stat. Phys.*, 1985, vol.38, pp.231-252.
40. S. S. Kim and D. N. Yoon, "Coarsening Behaviour of Mo Grains Dispersed in Liquid Matrix," *Acta Met.*, 1983, vol.31, pp.1151-1157.
41. G. C. Nicholson, "Grain Growth in Magnesium Oxide Containing a Liquid Phase," *J. Amer. Ceramic Soc.*, 1965, vol.48, pp.525-528.
42. R. Watanabe and Y. Masuda, "The Growth of Solid Particles in Fe-20 wt.% Cu Alloy During Sintering in the Presence of a Liquid Phase," *Trans. Japan Inst. Met.*, 1973, vol.14, pp.320-326.
43. S. S. Kang and D. N. Yoon, "Kinetics of Grain Coarsening During Sintering of Co-Cu and Fe-Cu Alloys with Low Liquid Contents," *Metall. Trans. A*, 1982, vol.13A, pp.1405-1411.
44. C. H. Kang and D. N. Yoon, "Coarsening of Cobalt Grains Dispersed in

Liquid Copper Matrix," *Metall. Trans. A*, 1981, vol.12A, pp.65-69.
45. A. Kannappan, "Particle Growth in Liquid Phase Sintering of W-Fe-Ni Alloys," International Team for Sintering, Report ITS-31, Ljubljana, Yugoslavia, 1971.
46. R. H. Krock, "Elastic and Plastic Deformation of Dispersed Phase Liquid Phase Sintered Tungsten Composite Materials," *Metals for the Space Age*, F. Benesovsky (ed.), Metallwerk Plansee, Reutte, Austria, 1965, pp.256-275.
47. T. K. Kang and D. N. Yoon, "Coarsening of Tungsten Grains in Liquid Nickel-Tungsten Matrix," *Metall. Trans. A*, 1978, vol.9A, pp.433-438.
48. R. Warren and M. B. Waldron, "Microstructural Development During the Liquid-Phase Sintering of Cemented Carbides. II. Carbide Grain Growth," *Powder Met.*, 1972, vol.15, pp.180-201.
49. R. Warren, "Microstructural Development During the Liquid-Phase Sintering of VC-Co Alloys," *J. Mater. Sci.*, 1972, vol.7, pp.1434-1442.
50. I. M. Stephenson and J. White, "Factors Controlling Microstructure and Grain Growth in Two-Phase and in Three-Phase Systems," *Trans. Brit. Ceramic Soc.*, 1967, vol.66, pp.443-483.
51. J. P. Sadocha and H. W. Kerr, "Grain-Ripening of the Lead-Rich Phase in Partially Molten Pb-Sb-Sn and Pb-Sb-Sn-As Solders," *Metal Sci. J.*, 1973, vol.7, pp.138-146.
52. R. D. McKellar and C. B. Alcock, "Particle Coarsening in Fused Salt Media," *Sintering and Catalysis*, G. C. Kuczynski (ed.), Plenum Press, New York, NY, 1975, pp.409-418.
53. P. J. Jorgensen and R. W. Bartlett, "Liquid-Phase Sintering of $SmCo_5$," *J. Appl. Phys.*, 1973, vol.44, pp.2876-2880.
54. L. Lindau and K. G. Stjernberg, "Grain Growth in TiC-Ni-Mo and TiC-Ni-W Cemented Carbides," *Powder Met.*, 1976, vol.19, pp.210-213.
55. H. E. Exner, "Qualitative and Quantitative Interpretation of Microstructures in Cemented Carbides," *Science of Hard Materials*, R. K. Viswanadham, D. J. Rowcliffe, and J. Gurland (eds.), Plenum Press, New York, NY, 1983, pp.233-262.
56. M. Fukuhara and H. Mitani, "Mechanisms of Grain Growth in Ti(C,N)-Ni Sintered Alloys," *Powder Met.*, 1982, vol.25, pp.62-68.
57. N. C. Kothari, "Densification and Grain Growth During Liquid-Phase Sintering of Tungsten-Nickel-Copper Alloys," *J. Less-Common Metals*, 1967, vol.13, pp.457-468.
58. D. S. Buist, B. Jackson, I. M. Stephenson, W. F. Ford, and J. White, "The Kinetics of Grain Growth in Two-Phase (Solid-Liquid) Systems," *Trans. British Ceram. Soc.*, 1965, vol.64, pp.173-209.
59. R. Watanabe and Y. Masuda, "The Growth of Solid Particles in Some Two-Phase Alloys During Sintering in the Presence of a Liquid Phase," *Sintering and Catalysis*, G. C. Kuczynski (ed.), Plenum Press, New York, NY, 1975, pp.389-398.
60. S. Takajo, W. A. Kaysser, and G. Petzow, "Analysis of Particle Growth by Coalescence During Liquid Phase Sintering," *Acta Met.*, 1984, vol.32, pp.107-113.
61. W. A. Kaysser, S. Takajo, and G. Petzow, "Particle Growth by Coalescence During Liquid Phase Sintering of Fe-Cu," *Sintering - Theory and Practice*, D. Kolar, S. Pejovnik, and M. M. Ristic (eds.), Elsevier Scientific, Amsterdam, Netherlands, 1982, pp.321-327.
62. E. G. Zukas, P. S. Z. Rogers, and R. S. Rogers, "Experimental Evidence for Several Spheroid Growth Mechanisms in the Liquid-Phase Sintered Tungsten-Base Composites," Los Alamos Scientific Laboratory,

Report No. LA-6223-MS, Los Alamos, NM, February 1976.
63. R. L. Fullman, "Measurement of Particle Size in Opaque Bodies," *Trans. TMS-AIME*, 1953, vol.197, pp.447-452.
64. S. S. Kang, S. T. Ahn, and D. N. Yoon, "Determination of Spherical Grain Size from the Average Area of Intersection in Ostwald Ripening," *Metallog.*, 1981, vol.14, pp.267-270.
65. M. Schreiner, T. Schmitt, E. Lassner, and B. Lux, "On the Origins of Discontinuous Grain Growth During Liquid Phase Sintering of WC-Co Cemented Carbides," *Powder Met. Inter.*, 1984, vol.16, pp.180-183.
66. K. Y. Eun, "The Abnormal Grain Growth and the Effects of Ni Substitution on Mechanical Properties in Sintered WC-Co Alloys," Ph.D. Thesis, Korea Advanced Institute of Science and Technology, Seoul, Korea, 1983.
67. H. E. Exner, E. Santa Marta, and G. Petzow, "Grain Growth in Liquid-Phase Sintering of Carbides," *Modern Developments in Powder Metallurgy*, vol.4, H. H. Hausner (ed.), Plenum Press, New York, 1971, pp.315-325.
68. E. Lassner, M. Schreiner, and B. Lux, "Influence of Trace Elements in Cemented Carbide Production: Part 2," *Inter. J. Refractory Hard Metals*, 1982, vol.1, pp.97-102.
69. B. Merideth and D. R. Milner, "The Liquid-Phase Sintering of Titanium Carbide," *Powder Met.*, 1976, vol.19, pp.162-170
70. D. Y. Kim and A. Accary, "Mechanisms of Grain Growth Inhibition During Sintering of WC-Co Based Hard Metals," *Sintering Processes*, G. C. Kuczynski (ed.), Plenum Press, New York, NY, 1980, pp.235-244.
71. M. F. Yan, R. M. Cannon, and H. K. Bowen, "Grain Boundary Migration in Ceramics," *Ceramic Microstructures '76*, R. M. Fulrath and J. A. Pask (eds.), Westview Press, Boulder, CO, 1977, pp.276-307.
72. R. Warren, "Effects of the Carbide Composition on the Microstructure of Cemented Binary-Carbides," *Planseeber. Pulvermetall.*, 1972, vol.20, pp.299-317.
73. L. H. Van Vlack and G. I. Madden, "Grain Growth Restrictions in Two-Phase Microstructures," *Trans. TMS-AIME*, 1964, vol.230, pp.1200-1202.
74. J. White, "Microstructure and Grain Growth in Ceramics in the Presence of a Liquid Phase," *Sintering and Related Phenomena*, G. C. Kuczynski (ed.), Plenum Press, New York, NY, 1973, pp.81-108.
75. O. K. Riegger, G. I. Madden, and L. H. Van Vlack, "The Microstructure of Periclase When Subjected to Steelmaking Variables," *Trans. TMS-AIME*, 1963, vol.227, pp.971-976.
76. T. H. Courtney, "Microstructural Evolution During Liquid Phase Sintering: Part II. Microstructural Coarsening," *Metall. Trans. A*, 1977, vol.8A, pp.685-689.
77. A. N. Niemi and T. H. Courtney, "Microstructural Development and Evolution in Liquid-Phase Sintered Fe-Cu Alloys," *J. Mater. Sci.*, 1981, vol.16, pp.226-236.

CHAPTER SEVEN

Special Treatments Involving Liquid Phases

A. Overview

The discussion up to now has focused on persistent liquid phase sintering. For the classic systems like cemented carbides and heavy alloys, liquid phase sintering is composed of three stages. During the entire sintering period there is a coexisting liquid which provides the total force for densification. Several alternatives exist to this traditional process. The advantages of these alternatives come from the improved rates of densification, and the unique compositions or microstructures possible. In examining these variants to traditional liquid phase sintering, the previously developed fundamentals are applied to describe the special techniques involving liquids during the sintering cycle.

B. Supersolidus Sintering

Supersolidus sintering is similar to traditional persistent liquid phase sintering; however, it is not as well understood. The major difference is the use of prealloyed powders instead of mixed powders. The sintering temperature is selected to be between the liquidus and solidus for the composition. At the sintering temperature, the liquid forms within each particle. As a consequence, each particle undergoes fragmentation and repacking, giving a homogeneous distribution of liquid. The resulting sintering rate is very rapid once the liquid forms. Figure 7.1 illustrates the steps associated with supersolidus sintering. In sequence, the steps are liquid formation, particle fragmentation, fragment rearrangement, grain repacking and sliding (a creep process), coarsening, and eventual pore elimination by solution-reprecipitation.

At the sintering temperature, a wide separation between the liquidus and solidus compositions is desirable (1). Even so, close temperature control is necessary to maintain adequate control over the microstructure and liquid content. In most instances the liquid wets the solid and the solid has a high solubility in the liquid. As discussed in earlier chapters, these factors are favorable for liquid phase sintering. In general, the uniform liquid formation throughout the microstructure gives homogeneous sintering, better than seen with mixed elemental powders.

The main applications for supersolidus sintering have been to high carbon steels, tool steels, nickel-base superalloys, and cobalt-base wear

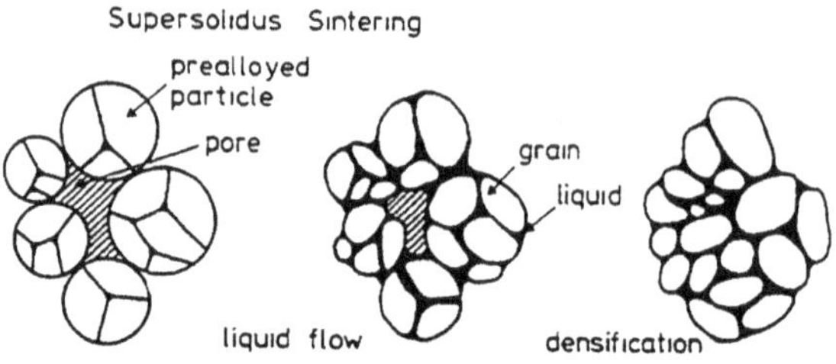

Figure 7.1 Supersolidus sintering involves the formation of a liquid along the grain boundaries in a prealloyed powder which leads to densification.

alloys. The key process control parameters are temperature and powder composition, since these two parameters dictate the volume fraction of liquid (2). The densification and sintering shrinkage increase as the amount of liquid increases (1,3,4). Figure 7.2 demonstrates the effect of temperature on the volume fraction of liquid and the sintered density for a loose packed nickel-base superalloy powder (4). These results demonstrate that high sintered densities are possible with at least 20 volume percent liquid. As a generalization, at least 15 to 20% liquid is necessary for significant densification by supersolidus sintering. For example, in a study of a Fe-0.9% C alloy by Lund and Bala (1), densification has been measured versus the volume fraction of liquid. The results of that study are shown in Figure 7.3. In spite of the short sintering time (10 minutes), considerable densification occurred with liquid contents over approximately 20%.

The results on several systems demonstrate the importance of composition and temperature as process control parameters. Temperature is important because of its effect on the volume fraction of liquid. Too high a sintering temperature leads to an excess quantity of liquid. In this case slumping of the compact is apparent and the microstructure exhibits extensive coarsening. Figure 7.4 demonstrates this effect for a nickel-based superalloy (5). The sintered density and grain size are shown as a function of the sintering temperature. The increasing volume fraction of liquid at the higher temperatures aids densification but also increases the sintered grain size. Eventually, excessive quantities of liquid lead to slumping. Alternatively, a low sintering temperature is unsatisfactory since there is insufficient liquid to wet the interparticle contacts. For the tool steels, temperatures must be controlled within 3°C of an optimal value for the composition. Temperature changes larger than this result in measurable differences in mechanical properties after sintering. Likewise, composition has a similar effect on the volume fraction of liquid at the sintering temperature. Kulkarni et al. (6) show that a relation between optimal sintering temperature and composition results from this interdependence. As a consequence, supersolidus sintering requires close temperature control to obtain full density, avoid shape distortion, and minimize microstructural coarsening (2,5,7-11).

Success in supersolidus sintering depends on a homogeneous liquid throughout the compact. Milling aids the initial sintering rate and additives can be mixed with the prealloyed powder to maintain compositional control

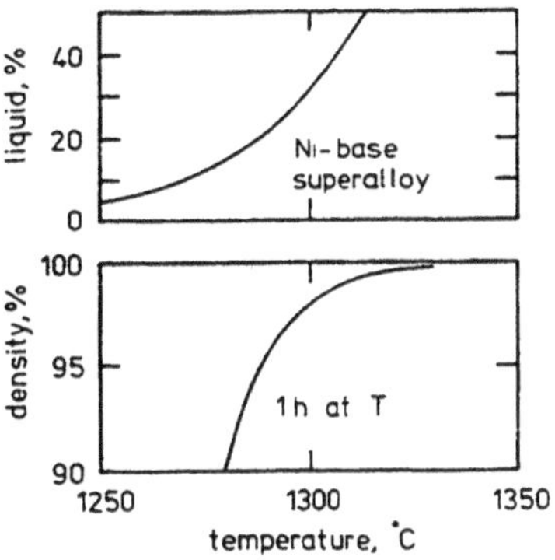

Figure 7.2 The amount of liquid versus supersolidus sintering temperature for a nickel-base superalloy and the corresponding density after 1 hour (4).

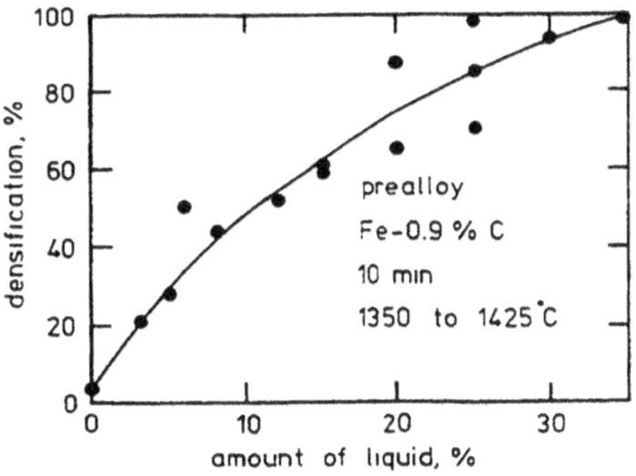

Figure 7.3 Densification for a Fe-C prealloyed powder sintered with various amounts of liquid as dictated by the sintering temperature (1).

during sintering (5,6,12). For low volume fractions of liquid, fine particles are more responsive; however, a wide particle size distribution is typically used to aid powder handling and packing. At high volume fractions of liquid, there is no significant effect from the initial particle size (3).

There are several problems with supersolidus sintering. Solid-state sintering during heating to the sintering temperature proves detrimental to final densification. The solid bonds inhibit rearrangement when the liquid forms. Additionally, the temperature at which optimal sintering is observed varies with the composition. Consequently, it proves difficult to simultaneously control both the temperature and variations in the powder composition. The rate of sintering is very rapid once the liquid forms. Although this is

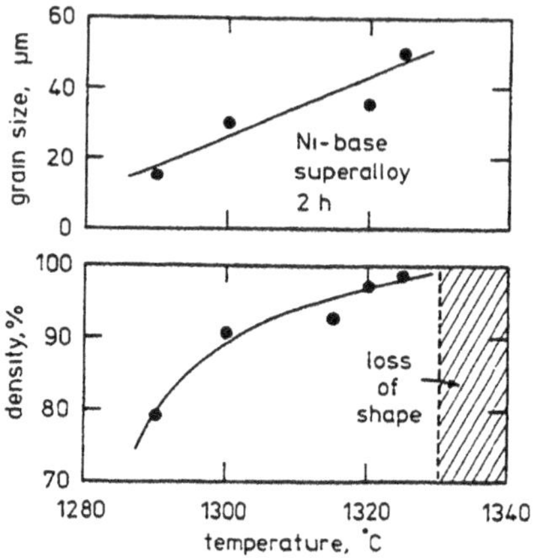

Figure 7.4 Grain size and density for a nickel-base superalloy shown as functions of the sintering temperature (which controls the volume of liquid) (5).

beneficial to obtaining high sintered densities, it presents a difficulty in controlling dimensions and the microstructure. Most successful applications of supersolidus sintering rely on vacuum to avoid trapped gas in the pores.

In spite of these recognized difficulties, the prospects for future use are excellent. Supersolidus sintering proves useful for full density processing of high alloy compositions. The newer sintering furnaces possess the temperature controls necessary for successful densification without distortion. Additionally, the process is understood in terms of the basic parametric influences, thus regression models are available to aid optimization. One major benefit is that supersolidus sintering is applicable to coarse powders. Also, it is useful for sintering rapidly solidified, high alloy, fine grain size materials to full density. This provides an opportunity for several future developments.

C. Infiltration

Infiltration is a two step variant of liquid phase sintering (13-21). Infiltration starts with a presintered rigid compact formed by solid-state sintering. The presintered compact provides a solid skeleton into which liquid is introduced during a second sintering treatment. The liquid is added to the compact (after presintering) from an external reservoir such as illustrated in Figure 7.5. Pore filling by the liquid relies on capillarity. A liquid with a low contact angle is drawn into the open pore structure of the presintered compact. Commonly, the liquid is formed from a preform which is melted on the surface of the solid skeleton. The resulting microstructure lacks pores and appears similar to other liquid phase sintered materials as shown in Figure 7.6. This figure is an optical micrograph of a Mo-Ag electrical contact formed by infiltration. Applications for this technique include Fe-Cu, Fe-B, TiC-Ni, Co-Cu, W-Ag, W-Cu, WC-Cu, WC-Ag, Mo-Ag, and Fe-C compositions

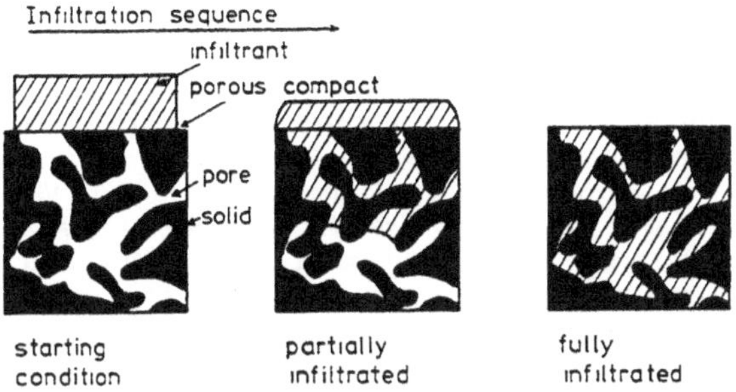

Figure 7.5 A view of the infiltration sequence where a molten metal forms on the surface of a porous compact and fills the pores based on capillarity.

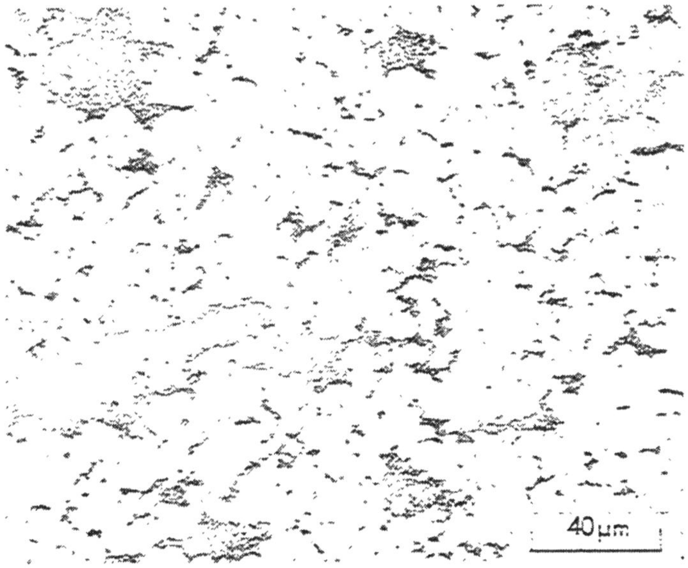

Figure 7.6 The microstructure of a Mo-Ag compact after infiltration with silver.

(22). It is the good dimensional control and the elimination of porosity which constitute the main attractions of infiltration.

Infiltration requires that the pore structure be open and interconnected; thus, the solid skeleton must have at least 10% porosity. The liquid must have a low viscosity and must wet the solid (23-25). Also, no intermediate compounds can form between the solid and liquid since they would block the infiltration path. Ideally, there should be little intersolubility between the solid and liquid. On the one hand, if the liquid is soluble in the solid, then it will be transient and inhibit infiltration because of diffusional solidification. Alternatively, if the solid has a high solubility in the liquid, then the compact will undergo erosion, slumping, and coarsening. Furthermore, to preserve rigidity during infiltration, the dihedral angle should be greater than zero. To prevent swelling and surface erosion by the liquid dissolving the solid as it enters the compact, it is common to use a saturated liquid composition. Figure 7.7 shows the time dependent swelling of porous iron infiltrated with pure copper and an alloy of copper and iron (26). Infiltration with pure copper gives greater attack of the solid grain boundaries and produces greater swelling. The presaturated liquid does not give as much penetration of the iron grain boundaries, hence there is less expansion.

The infiltration process occurs rapidly with pore filling, grain dissolution, and fragmentation taking place simultaneously. The initial depth of infiltration h in a porous compact has been estimated as follows (15,24):

$$h = (2/\pi)\, [r\, t\, \gamma \cos\theta/(2\, \mu)]^{1/2} \qquad (7.1)$$

where r is the pore radius, γ is the liquid-vapor surface energy, t is the time, θ is the contact angle between the solid and liquid, and μ is the viscosity. The pore radius is approximately 25% of the mean grain separation. Equation (7.1) shows that the infiltrated pore length increases with the square-root of time. In typical applications, infiltration is complete in 2 to 5 minutes. The infiltration rate is thermally activated; thus, it increases with an exponential dependence on temperature (27). With prolonged exposure of the solid to the liquid, grain growth will result. The rate of grain growth is the same as that noted during the final stage of liquid phase sintering; the grain size increases with time to approximately the 1/3 power (14,16). As a consequence, the time at temperature after infiltration is kept short.

There are some common problems with infiltration (17,18). The process is very sensitive to surface contamination, leading to the use of various surface cleaning treatments to aid wetting (28). Craters are often seen on surfaces exposed to the infiltrating liquid. These craters are due to erosion; the selective dissolution of solid to attain solubility equilibrium in the liquid during ingress of the liquid. On cooling, the liquid may alter the sintered microstructure in a way which is cooling rate sensitive. Swelling is common, with the magnitude depending on the solid skeleton porosity, grain size, and dihedral angle. Swelling originates from grain separation due to a penetrating liquid. Typically, the best mechanical properties are observed when the volume of infiltrant is slightly less than the pore volume. Excessive quantities of liquid will cause slumping and separation of the solid skeleton. Alternatively, insufficient liquid results in residual pores.

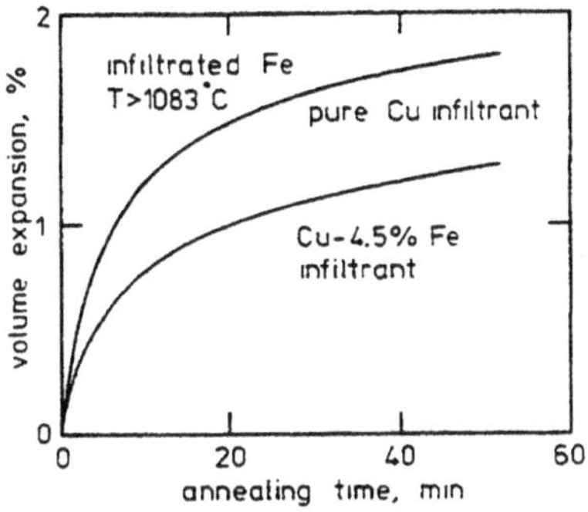

Figure 7.7 The volumetric expansion associated with grain boundary penetration by the infiltrant. The use of a prealloyed Cu-Fe infiltrant reduces swelling (26).

D. Pressure Assisted Densification

An external pressure during liquid phase sintering will aid densification and pore collapse (29-36). The liquid can not sustain a shear stress; thus, it lubricates particle sliding as induced by an external stress. As a consequence, an external stress eliminates pores and inhomogeneous regions which are otherwise stable in some liquid phase sintered materials. Such a technique is of great value in densifying systems with poor wetting or unstable compounds. For example, cemented diamond composites require a high temperature and high pressure to attain densification while preventing decomposition of the diamond (37). Pressure assisted densification has several variants which share the common feature of an external stress and a low volume fraction of liquid.

With an external stress, rearrangement processes play a larger role in densification. In the rearrangement stage, the rate of densification increases as the capillary force increases. An external stress directly supplements the capillary force (31,32,36). Depending on the particle size, the applied stress can make a significant increase in the capillary force. Thus, the amount of initial stage rearrangement initially is increased by an applied stress. Kingery et al. (32) treat the case of solution-reprecipitation controlled densification with an external stress. The external stress is additive to the capillary force, giving a shrinkage $\Delta L/L_o$ dependence as follows:

$$(\Delta L/L_o)^3 = [\, g\ \delta\ D\ C\ \Omega\ t/(G^3\ k\ T)](\sigma + 2\ \gamma/r) \tag{7.2}$$

where g is a geometric constant, δ is the thickness of the intergranular film of liquid, D is the diffusion rate in the liquid, C is the solid concentration in the liquid, Ω is the atomic volume, t is the sintering time, G is the grain size, k is Boltzmann's constant, T is the absolute temperature, γ is the liquid-vapor surface energy, r is the pore radius, and σ is the applied stress. As densification takes place, the pore volume decreases and the surface energy term in Equation (7.2) diminishes. In the final stage where densification is controlled by diffusion, creep processes can be induced by the external stress. For creep controlled densification, the densification rate will directly depend on the stress. In the final stage of sintering, Wang and Raj (33) suggest the shrinkage rate $d(\Delta L/L_o)/dt$ will vary as follows:

$$d(\Delta L/L_o)/dt = B\,\sigma^n \exp(-Q/(k\,T))/G^3 \qquad (7.3)$$

where B is a collection of material constants, Q is the activation energy for diffusion, and G is the grain size. According to Equation (7.3), the rate of diffusion controlled shrinkage depends on the applied stress. In several studies, the stress exponent n has been found to be near unity.

Stress aids densification and has an effect similar to increasing the amount of liquid. Figure 7.8 shows the porosity for Cu-Bi compacts with two different external stresses (32). As the external stress increases, the pore volume decreases. Figure 7.8 further illustrates this effect by contrasting the behavior of copper with and without a liquid. The presence of a liquid or an increase in the applied pressure results in a higher sintered density. Figure 7.9 also shows the pressure effect using data for a nickel-base superalloy (29). As the pressure increased, the shrinkage also increased. Note that these are relatively small external stresses, up to 4 atmospheres, giving significant changes in the supersolidus sintering behavior. Alternatively, Figure 7.10 demonstrates the volume fraction of liquid effect under a constant pressure (38). This figure gives the porosity dependence on the amount of liquid as formed by yttria additives to silicon nitride. At low volumes of liquid, the solid-solid contacts give rigidity to the compact and offset the applied pressure effect. As the quantity of liquid is raised, then densification is improved. The effect of pressure is more pronounced with coarser particle sizes since the capillary force depends on the inverse of the grain size. Thus, as shown in Figure 7.11 an increase in the particle (grain) size gives a greater relative effect from the applied stress (31). In Figure 7.11, the coarser diamond powder shows the least sintering shrinkage with no external stress. As the pressure increases the particle size effect becomes smaller. Although the densification rate is improved by an external pressure, the basic microstructural relations are not altered.

E. Transient Liquids

An interesting variant to traditional liquid phase sintering involves a transient liquid which solidifies by diffusional homogenization during sintering (39-66). The transient liquid forms between mixed ingredients during heating to the sintering temperature. Figure 7.12 presents the phase diagrams of two example systems which could be processed using a transient liquid. The solid does not change crystal structure due to alloying with the liquid. For the first case shown in Figure 7.12, the liquid forms from a eutectic reaction between the mixed constituents. Alternatively for the second case, the liquid forms from the mixture of A (additive) and B (base) when the additive melts. Unlike persistent liquid phase sintering, the liquid has a high solubility in the solid and disappears with sintering time. The benefits of transient liquid phase sintering are easy compaction of elemental powders (as opposed to prealloyed powders) and excellent sintering without the coarsening difficulties associated with a persistent liquid. However, because the liquid content

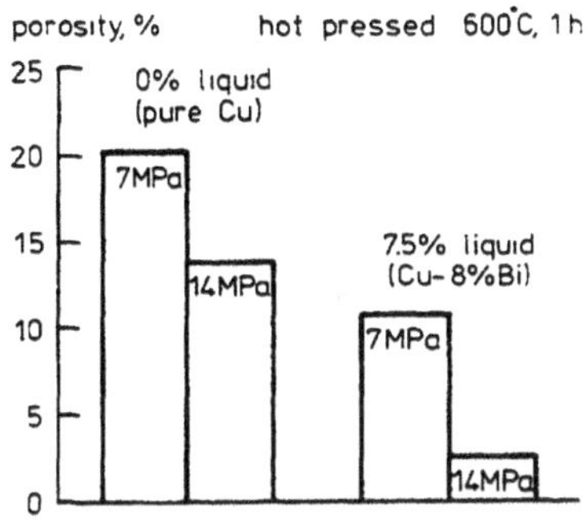

Figure 7.8 The porosity after hot pressing at 600°C for pure Cu (0% liquid) and Cu-Bi (7.5% liquid) compacts, using two different compaction pressures (32).

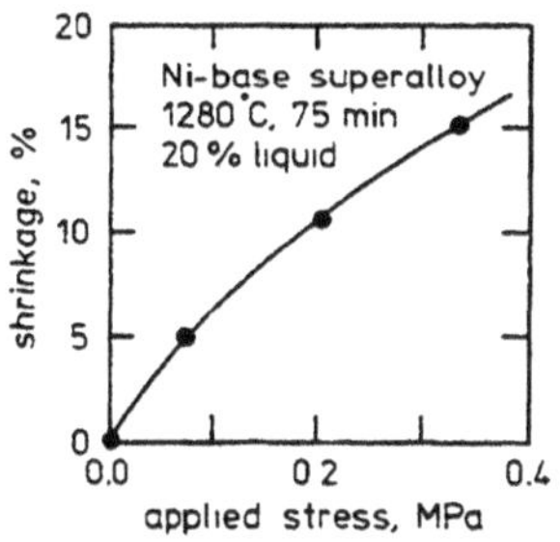

Figure 7.9 Shrinkage versus applied pressure during pressure assisted densification of a superalloy powder containing 20% liquid (29).

depends on several processing parameters, transient liquid phase sintering is highly sensitive to processing conditions. Additionally, the solubility ratio necessary for the liquid to be transient induces swelling during heating to the sintering temperature. For applications such as porous bronze bearings this is beneficial, while for structural materials the porosity and swelling are detrimental.

Several application exist for transient liquid phase sintering. A common application is as a dental amalgam based on silver and mercury. Here solid silver-based alloy powder is mixed with liquid mercury and the slurry is compressed into a dental cavity prior to solidification of the liquid. There are several variants to the basic amalgamation reaction, as reviewed by Jangg (67). Another common application is in forming porous bronze bearings (self-lubricating) from mixed Cu-Sn powders (68-70). As shown by the micrographs of Figure 7.13, the tin melts and spreads into the copper structure forming a large pore. Subsequent diffusional homogenization is rapid,

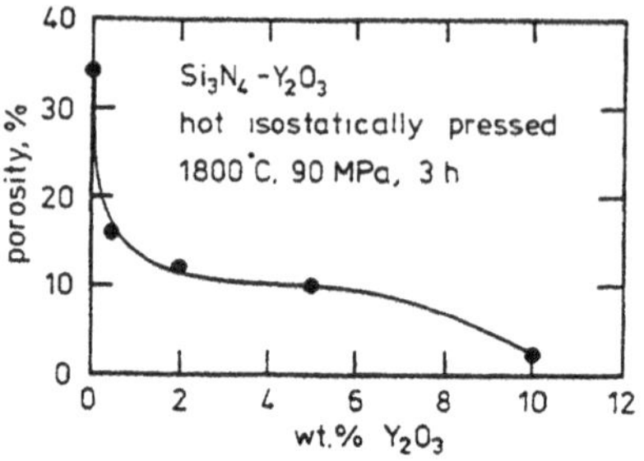

Figure 7.10 Porosity of silicon nitride with yttria additions which produce a liquid. The compacts were formed by hot isostatic pressing at 1800°C (38).

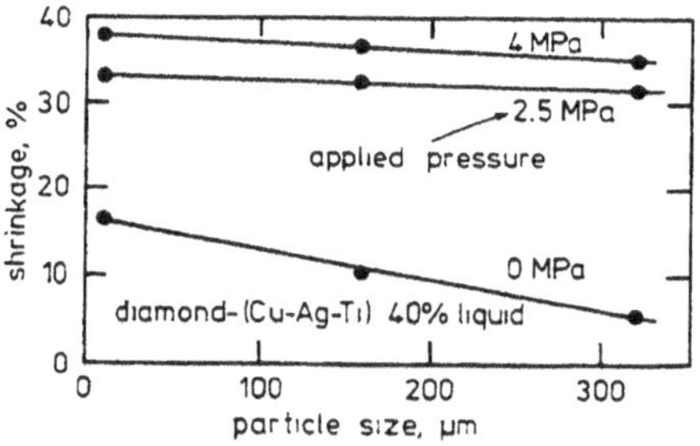

Figure 7.11 The effect of the diamond particle size on shrinkage during hot pressing with three different applied pressures for compacts with 40% liquid (31).

leading to a solid alloy structure with large pores. Other applications include structural ferrous alloys, copper alloys, magnetic materials, and alumina-based ceramics. These systems exhibit diffusional interactions between the constituents during heating to the liquid formation temperature. This interaction affects the subsequent liquid phase behavior and swelling during heating.

The requirements for transient liquid phase sintering include mutual intersolubility between the components with the final composition existing within a single phase region. Furthermore, the liquid must wet the solid and give a high diffusion rate for the solid. Under these conditions rapid sintering is anticipated when the liquid forms. Generally, the observed steps are as follows (54,56,71): 1) swelling by interdiffusion prior to melt formation (Kirkendall porosity), 2) melt formation, 3) spreading of the melt and generation of pores at prior additive particle sites, 4) melt penetration along solid-solid contacts, 5) rearrangement of the solid grains, 6) solution-reprecipitation induced densification, 7) diffusional homogenization, 8) loss of melt, and 9) formation of a rigid solid structure. The actual steps depend on the

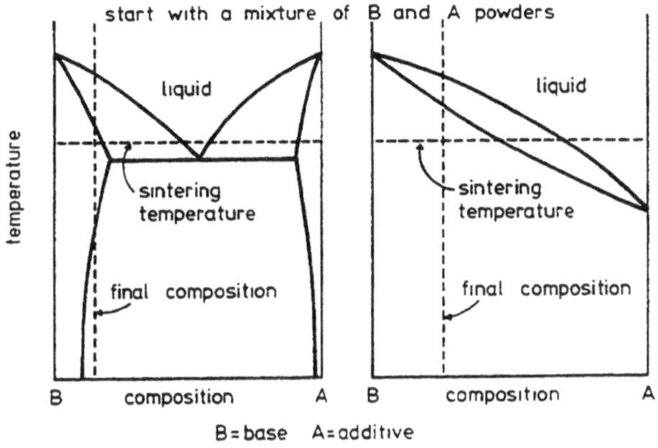

Figure 7.12 Two binary phase diagrams which could form the basis for transient liquid phase sintering using the indicated compositions and temperatures.

several process variables, including particle sizes, amount of additive, heating rate, and maximum temperature. Swelling is common during heating because of the required solubility of additive in the base, especially if intermediate compounds form between the constituents (71,72). The densification associated with transient liquid phase sintering depends on the amount of liquid formed and the length of time the liquid exists. Too much liquid can generate large pores due to spreading and penetration of the melt away from the additive particle sites. Figure 7.14 shows this liquid spreading away from the additive particle site for a Fe-Ti compact. A pore has formed at the prior Ti particle site due to the outward diffusion of Ti. The liquid is penetrating the neighboring solid grain boundaries, which leads to resolidification after separating the solid grains (61). Typically, the liquid lasts for only a few minutes, thus pore refilling does not occur and large pores result. However, if the liquid persists, then densification will follow. Alternatively, with too little liquid minimal densification is observed.

Heating rate is important to transient liquid phase sintering. Figure 7.15 illustrates the effect of heating rate on homogeneity, amount of liquid, and dimensional change during transient liquid phase sintering. During heating, diffusional homogenization leads to pore formation by a Kirkendall effect. The amount of swelling increases with the amount of interdiffusion; thus, more swelling is seen with slower heating rates. Furthermore, diffusional homogenization reduces the amount of liquid formed at the eutectic temperature and the degree of sintering. After reaching the sintering temperature, the amount of liquid and the length of time the liquid exists determine the net shrinkage. The amount of liquid depends on the degree of homogenization and the initial additive concentration. A relation between the volume fraction of liquid V_L, additive concentration C, and heating rate χ follows:

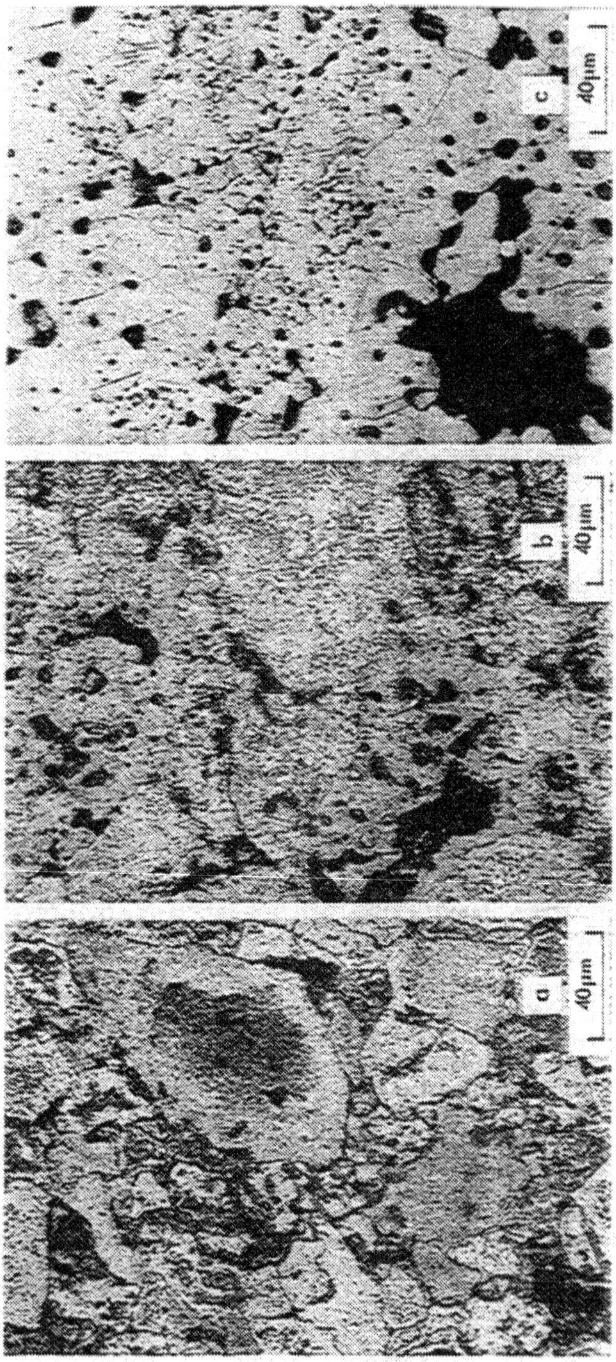

Figure 7.13 Pore formation in the microstructure of Cu-10% Sn mixed powder compacts heated to (a) 280°C, (b) 600°C, and (c) 810°C.

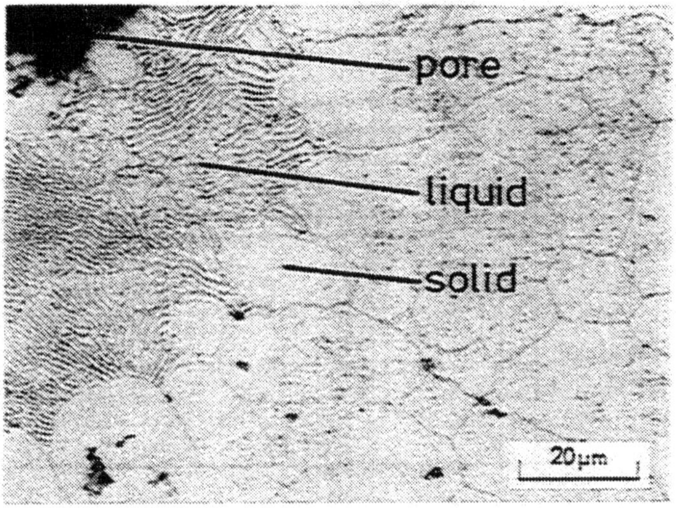

Figure 7.14 Melt penetration of solid grain boundaries in a Fe-Ti powder compact, showing pore formation at the prior Ti particle site (photo courtesy J. Dunlap).

$$1 - [V_L/(f\ C)]^{1/3} = \kappa\ (T_L/x)^{1/2} \tag{7.4}$$

where f and κ are proportionality constants, and T_L is the liquid formation temperature.

According to Equation (7.4), more liquid is formed with faster heating rates and larger quantities of liquid forming additive. The amount of liquid is reduced by use of finer particle sizes. Depending on the phase diagram, solid-state interdiffusion can give an intermediate compound as a surrounding envelope around the additive particles as illustrated in Figure 7.16. This figure is a scanning electron micrograph of a Fe-Al powder compact heated to 635°C, prior to the aluminum melting. The electron contrast clearly shows the intermetallic envelope formed around the aluminum particle by solid-state diffusion. This envelope can inhibit subsequent interdiffusion (55). The thickness w of the intermediate compound initially increases with time t as follows (41):

$$w \sim (D\ t)^{1/2} \tag{7.5}$$

where D is the controlling diffusion rate in the compound layer. The growth of the compound typically results in compact swelling and may result in an exothermic reaction which heats the compact. Often the swelling reaction is most intense just prior to the liquid formation temperature (57) and is not

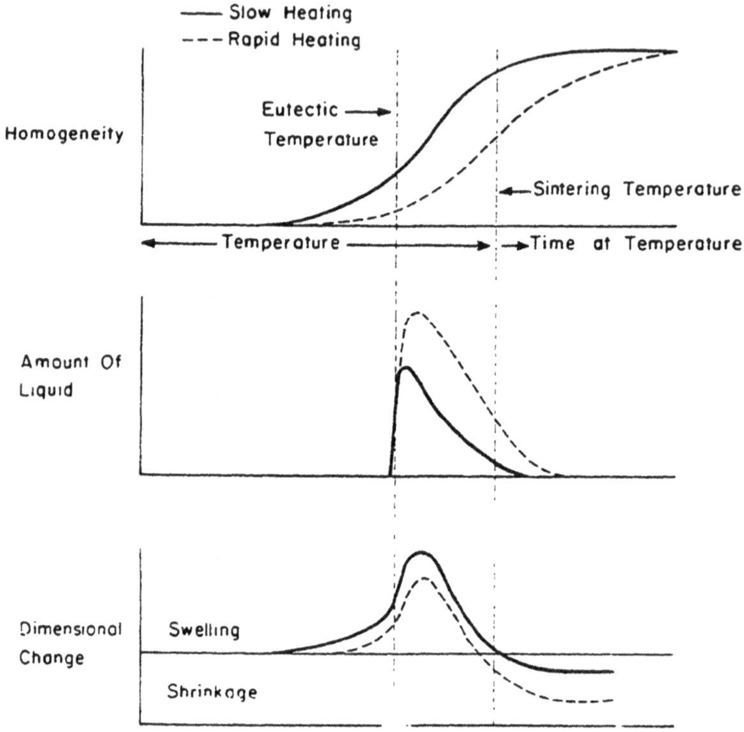

Figure 7.15 An example of the heating rate effect on transient liquid phase sintering; slow heating gives more diffusional homogenization, less liquid, and more swelling.

due to trapped gas in the pores (42). The amount of swelling during heating depends on the heating rate and additive content, which control the degree of interdiffusion (63). These factors are demonstrated in Figure 7.17, showing the swelling variations with processing conditions for Fe-Al compacts. The amount of densification after liquid formation depends on how much liquid was formed and the length of time the liquid persists. In turn the heating rate, additive content, and particle size are dominant with respect to densification (45,59,64). Figure 7.18 shows the interrelation between titanium content and heating rate for sintered Fe-Ti powder mixtures after heating to 1350°C for 1 hour. The combinations giving optimal densification are shown along with the model fit using Equation (7.4). Generally faster heating rates allow use of smaller particle sizes and lower additive contents, and give superior sintered properties.

Generally time at the sintering temperature is not a major parameter. The compact viscosity increases with time because of a decreasing quantity of liquid. Thus, the densification process slows rapidly (41,58).

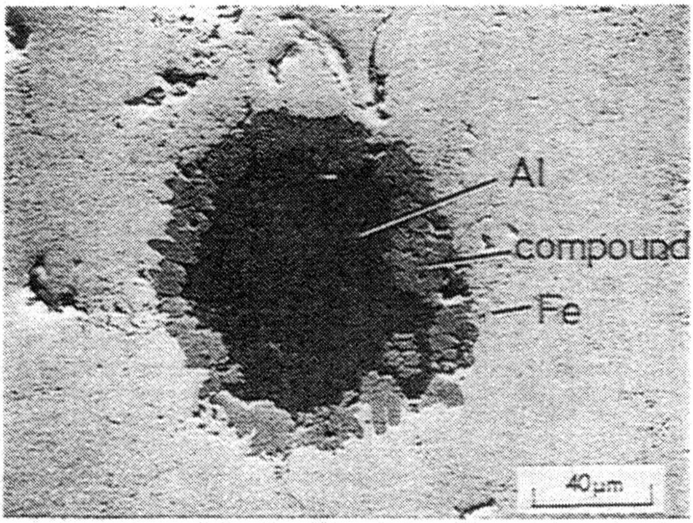

Figure 7.16 A scanning electron micrograph of the intermetallic envelope formed around a Al particle in a Fe-Al powder compact heated to 635°C (photo courtesy D. J. Lee).

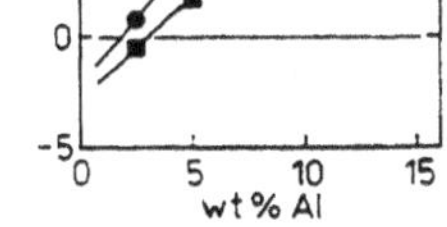

Figure 7.17 Swelling versus the amount of aluminum in Fe-Al compacts sintered at 1250°C, 1 hour with two different heating rates (71).

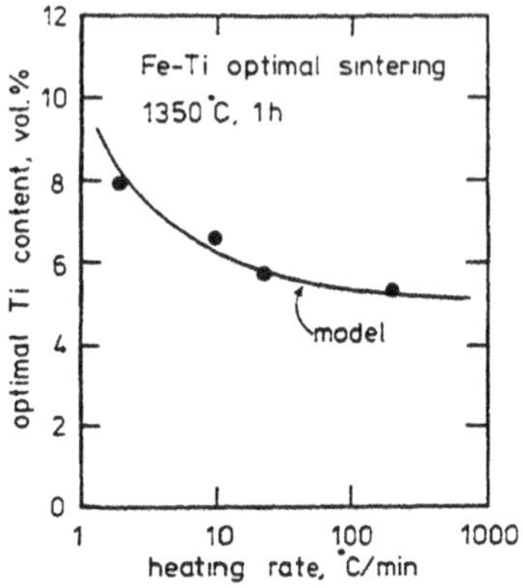

Figure 7.18 In Fe-Ti powder compacts processed by transient liquid phase sintering, optimal densification depends on the heating rate and the amount of Ti (55).

As expected with a complex processing technique, there are several potential problem areas. Pore formation during heating is a major difficulty. These pores can become stabilized and remain during the balance of the sintering cycle. In some cases, the swelling process associated with the high additive solubility in the base can generate stresses and crack the compact. Overall, transient liquid phase sintering is more sensitive to process variables than other forms of sintering. Consequently, optimal sintering requires a balance between the processing parameters. Only in a few studies have these interrelations been developed and processing optimized. Successful applications of transient liquid phase sintering depend on striking such balances for optimal sintered properties.

A novel combination of transient liquid phase sintering and infiltration processes has been described by Langford and Cunningham (73) as diffusional solidification. In this case, packed powders are heated prior to infiltration with a liquid. The overall alloy composition is selected to be in a single phase solid region of the phase diagram at the infiltration temperature. Thus, the liquid is rich in solute compared to the solid. After infiltration, the solute diffuses into the solid to homogenize the system and causes the liquid to solidify. One advantage of this approach is that it avoids the gross porosity observed in some cases of transient liquid phase sintering.

F. Reactive Sintering

Reactive sintering is similar to transient liquid phase sintering. It is characterized by a large heat liberation due to a reaction between the constituent powders (74). Rapid sintering results from the formation of a liquid (possibly due to reactive self-heating from an exothermic reaction) between the constituent powders. Figure 7.19 illustrates one type of phase diagram anticipated for reactive sintering. In this case the process starts with pure components A and B to form the compound AB. By appropriate temperature and composition selection a liquid is formed during the reaction. A sequence of steps expected for reactive liquid phase sintering is shown in Figure 7.20.

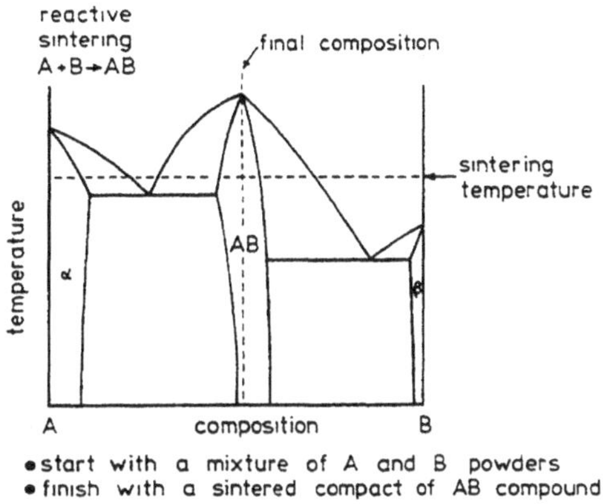

Figure 7.19 A binary phase diagram showing the characteristics favorable for reactive liquid phase sintering of the AB compound from mixed A and B powders.

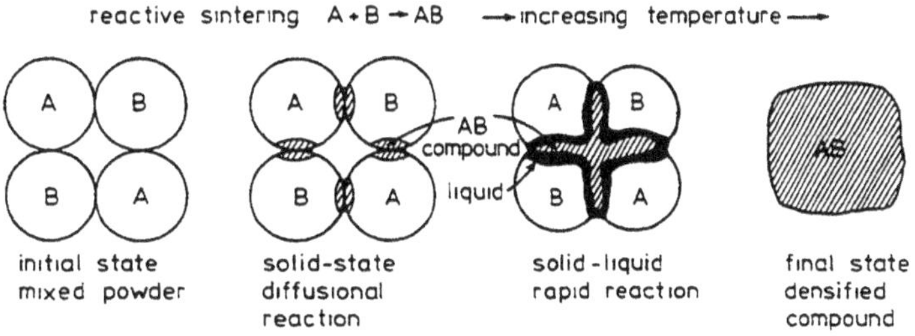

Figure 7.20 A typical sequence of events expected in reactive sintering involving the formation of a liquid from a mixture of A and B powders.

This figure corresponds to an increasing degree of reaction in going from the left to the right. The initial compact is composed of mixed powders which diffusionally interact to form the compound during heating. When a liquid forms in the compact, rapid compound formation begins, with liquid flow into the pores. The final product is typically a compound (as opposed to a solid solution with transient liquid phase sintering). The compound is a single phase material which has densified due to simultaneous sintering and reaction (72,75-80).

An example of reactive sintering is found in the synthesis of compounds like titanium-carbide or titanium-nitride from the constituent powders. In the study by Quinn and Kohlstedt (75), substoichiometric TiC was reacted with Ti powder to form TiC. A eutectic liquid formed during the sintering treatment

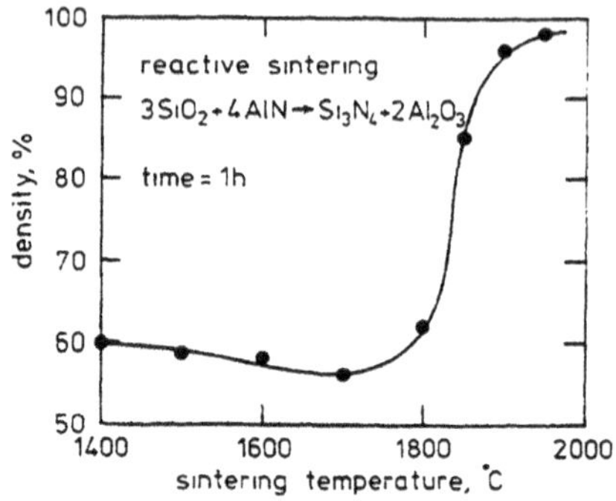

Figure 7.21 The sintered density of Si-Al-O-N compounds versus the temperature. At temperatures above 1800°C a liquid aids densification (76).

which consumed the excess titanium. During the reaction, grain growth and densification were observed due to rapid diffusion rates in the liquid. However, in spite of helpful chemical reactions, pore formation is common in reactive sintering systems. Densification is rapid once the liquid forms and control of the liquid duration depends on control of the temperature, particle size, and green density.

Figure 7.21 shows the densification data for a variant of reactive sintering (76). In this case silica and aluminum nitride are reacted to form silicon nitride and alumina at various temperatures. The sintered density decreases with increasing temperature up to 1800°C where the first liquid forms. At temperatures over 1800°C high sintered densities are observed using this technique.

Reactive sintering is still in the developmental stage. However, studies to date have identified several processing sensitivities and problems. Swelling due to the short liquid duration time is a major problem. The exothermic reaction must be controlled to maintain dimensions and avoid microstructure damage. Because of the limited knowledge on process control, most of the reactive sintering is being applied to forming compounds for subsequent consolidation. However, the potential exists for direct consolidation of materials into useful engineering shapes by reactive sintering, especially when combined with an external pressure (79,81).

G. Summary

This chapter has extended the treatment of liquid phase sintering to include several nontraditional forms of processing. In each case there is a common feature of coexisting liquid and solid phases. The source of the liquid, its duration, or the nature of the driving force for pore elimination differ from that treated by the classical theory of liquid phase sintering. It has been demonstrated that these differences typically complicate the processing; however, the fundamental considerations involved in sintering are similar. Good wetting of the solid by the liquid is necessary for rapid densification. A uniform liquid increases both the magnitude and the rate of sintering. In general, excessive quantities of liquid are detrimental to the sintered microstructure and properties. These variants on traditional liquid phase sintering are useful in several systems. In this chapter, it has been demonstrate that the main process variables can be understood in terms of the basic features considered in persistent liquid phase sintering.

H. References

1. J. A. Lund and S. R. Bala, "Supersolidus Sintering," *Modern Developments in Powder Metallurgy*, vol.6, H. H. Hausner and W. E. Smith (eds.), Metal Powder Industries Federation, Princeton, NJ, 1974, pp.409-421.
2. J. A. Lund, R. G. Butters, and C. H. Weaver, "Supersolidus Sintering of Prealloyed Monel Powder," *Powder Met. Inter.*, 1972, vol.4, pp.173-174.
3. C. Guyard, C. H. Allibert, J. Driole, and G. Raisson, "Liquid Phase Sintering of Prealloyed Powders of Co-Base Alloy," *Sci. Sintering*, 1981, vol.13, pp.149-163.
4. M. Jeandin, J. L. Koutny, and Y. Bienvenu, "Liquid Phase Sintering of Nickel Base Superalloys," *Powder Met.*, 1983, vol.26, pp.17-22.
5. R. Kieffer, G. Jangg, and P. Ettmayer, "Sintered Superalloys," *Powder Met. Inter.*, 1975, vol.7, pp.126-130.
6. K. M. Kulkarni, A. Ashurst, and M. Svilar, "Role of Additives in Full Dense Sintering of Tool Steels," *Modern Developments in Powder Metallurgy*, vol.13, H. H. Hausner, H. W. Antes and G. D. Smith (eds.), Metal Powder Industries Federation, Princeton, NJ, 1981, pp.93-120.
7. P. J. McGinn, A. E. Miller, P. Kumar, and A. J. Hickl, "Mechanisms of Liquid Phase Sintering in Stellite Alloy No. 6 PM," *Prog. Powder Met.*, 1982, vol.38, pp.449-462.
8. M. T. Podob, L. K. Woods, P. Beiss, and W. Huppmann, "The Mechanism of Sintering High Speed Steel to Full Density," *Modern Developments in Powder Metallurgy*, vol.13, H. H. Hausner, H. W. Antes and G. D. Smith (eds.), Metal Powder Industries Federation, Princeton, NJ, 1981, pp.71-92.
9. L. Cambal and J. A. Lund, "Supersolidus Sintering of Loose Steel Powders," *Inter. J. Powder Met.*, 1972, vol.8, pp.131-140.
10. R. Wahling, P. Beiss, and W. J. Huppmann, "Sintering Behaviour and Performance Data of HSS-Components," *Proceedings Sintering Theory and Practice Conference*, The Metals Society, London, UK, 1984, pp.15.1-15.5.
11. K. Hajmrle and R. Angers, "Sintering of Inconel 718," *Inter. J. Powder Met. Powder Tech.*, 1980, vol.16, pp.255-266, and 412.
12. P. J. McGinn, P. Kumar, A. E. Miller, and A. J. Hickl, "Carbide Composition Change During Liquid Phase Sintering of a Wear Resistant Alloy," *Metall. Trans. A*, 1984, vol.15A, pp.1099-1102.
13. P. Schwarzkopf, "The Mechanism of Infiltration," *Symposium on Powder Metallurgy*, Special Report 58, The Iron and Steel Institute, London, UK, 1956, pp.55-58.
14. L. P. Skolnick, "Grain Growth of Titanium Carbide in Nickel," *Trans. TMS-AIME*, 1957, vol.209, pp.438-442.
15. S. Banerjee, R. Oberacker, and C. G. Goetzel, "Mechanism of

Capillary-Force Induced Infiltration of Iron Skeletons with Cast Iron," *Inter. J. Powder Met. Powder Tech.*, 1984, vol.20, pp.325-341.

16. R. Savage and V. A. Tracey, "The Fabrication of Two Phase Alloys in the Ruthenium-Gold-Palladium System," *Modern Developments in Powder Metallurgy*, vol.5, H. H. Hausner (ed.), Plenum Press, New York, NY, 1971, pp.273-286.
17. K. A. Thorsen, S. Hansen, and O. Kjaergaard, "Infiltration of Sintered Steel with a Near Eutectic Fe-C-P Alloy," *Powder Met. Inter.*, 1983, vol.15, pp.91-93.
18. A. K. Mashkov, V. V. Chernienko, and Z. P. Gutkovskaya, "Development of a Process for the Production of Dense Sintered Materials by the Method of Infiltration of Porous Blanks With Low-Melting-Point Iron-Boride Alloys," *Soviet Powder Met. Metal Ceram.*, 1973, vol.12, pp.32-36.
19. P. W. Taubenblat and W. E. Smith, "Infiltration of Iron P/M Parts," *Copper Base Powder Metallurgy*, P. W. Taubenblat (ed.), Metal Powder Industries Federation, Princeton, NJ, 1980, pp.111-121.
20. A. N. Ashurst, E. Klar, and H. R. McCurdy, "Copper Infiltration of Steel: Properties and Applications," *Prog. Powder Met.*, 1983, vol.39, pp.163-182.
21. E. R. Lumpkins, Jr., "A Theoretical Review of the Copper Infiltration of PM Components," *Powder Met. Inter.*, 1985, vol.17, pp.120-123.
22. F. V. Lenel, *Powder Metallurgy Principles and Applications*, Metal Powder Industries Federation, Princeton, NJ, 1980.
23. A. J. Shaler, "Theoretical Aspects of the Infiltration of Powder Metallurgy Products," *Inter. J. Powder Met.*, 1965, vol.1, no.1, pp.3-14.
24. K. A. Semlak and F. N. Rhines, "The Rate of Infiltration of Metals," *Trans. TMS-AIME*, 1958, vol.212, pp.325-331.
25. A. F. Lisovskii, "Thermodynamics of the Penetration of a Liquid Phase into Sintered Composites," *Soviet Powder Met. Metal Ceram.*, 1974, vol.13, pp.832-835.
26. D. Berner, H. E. Exner, and G. Petzow, "Swelling of Iron-Copper Mixtures During Sintering and Infiltration," *Modern Developments in Powder Metallurgy*, vol.6, H. H. Hausner and W. E. Smith (eds.), Metal Powder Industries Federation, Princeton, NJ, 1974, pp.237-250.
27. A. A. Kurilko, G. A. Kurshev, V. A. Rudyuk, and Y. V. Naidich, "Kinetic Laws of Infiltration of Porous Titanium by Lead and Indium Melts," *Soviet Powder Met. Metal Ceram.*, 1984, vol.23, pp.686-689.
28. C. G. Goetzel and A. J. Shaler, "Mechanism of Infiltration of Porous Powder Metallurgy Parts," *J. Metals*, 1964, vol.16, pp.901-905.
29. M. Jeandin, J. L. Koutny, and Y. Bienvenus, "Rheology of Solid-Liquid P/M Astroloy - Application to Supersolidus Hot Pressing of P/M Superalloys," *Inter. J. Powder Met. Powder Tech.*, 1982, vol.18, pp.217-223.
30. Y. V. Naidich, I. A. Lavrinenko, and V. A. Evdokimov, "Liquid Phase Sintering under Pressure of Tungsten-Nickel-Copper Composites," *Soviet Powder Met. Metal Ceram.*, 1977, vol.16, pp.276-280.
31. Y. V. Naidich, I. A. Lavrienko, and V. A. Evdokinov, "Densification During Liquid-Phase Sintering in Diamond-Metal Systems. 2. Sintering under Pressure," *Soviet Powder Met. Metal Ceram.*, 1974, vol.13, pp.113-117.
32. W. D. Kingery, J. M. Woulbroun, and F. R. Charvat, "Effects of Applied Pressure on Densification During Sintering in the Presence of a

Liquid Phase," *J. Amer. Ceramic Soc.*, 1963, vol.46, pp.391-395.
33. J. G. Wang and R. Raj, "Mechanisms of Superplastic Flow in a Fine-Grained Ceramic Containing some Liquid Phase," *J. Amer. Ceramic Soc.*, 1984, vol.67, pp.399-409.
34. R. M. Spriggs and S. K. Dutta, "Pressure Sintering - Recent Advances in Mechanisms and Technology," *Sci. Sintering*, 1974, vol.6, pp.1-24.
35. S. R. Jurewicz and E. B. Watson, "Distribution of Partial Melt in a Felsic System: The Importance of Surface Energy," *Contrib. Mineral. Pwtrol.*, 1984, vol.85, pp.25-29.
36. Y. V. Naidich, I. A. Lavrinenko, and V. A. Evdokimov, "Densification in Liquid Phase Sintering under Pressure in the System Tungsten-Copper," *Soviet Powder Met. Metal Ceram.*, 1974, vol.13, pp.26-30.
37. R. H. Wentorf, R. C. DeVries, and F. P. Bundy, "Sintered Superhard Materials," *Science*, 1980, vol.208, pp.873-880.
38. O. Yeheskel, Y. Gefen, and M. Talianker, "Hot Isostatic Pressing of Silicon Nitride with Yttria Additions," *J. Mater Sci.*, 1984, vol.19, pp.745-752.
39. P. Lindskog, "The Effect of Phosphorus Additions on the Tensile, Fatigue, and Impact Strength of Sintered Steels Based on Sponge Iron," *Powder Met.*, 1973, vol.16, pp.374-386.
40. B. Rieger, W. Schatt, and C. Sauer, "Combined Mechanical Activation and Sintering with a Short-Time Occurance of a Liquid Phase," *Inter. J. Powder Met. Powder Tech.*, 1983, vol.19, pp.29-41.
41. L. Albano-Muller, F. Thummler, and G. Zapf, "High-Strength Sintered Iron-Base Alloys by using Transition Metal Carbides," *Powder Met.*, 1973, vol.16, pp.236-256.
42. K. V. Savitskii, V. I. Itin, Y. I. Kozlov, and A. P. Savitskii, "The Effect of the Particle Size of Aluminum Powder on the Sintering of a Cu-Al Alloy in the Presence of a Liquid Phase," *Soviet Powder Met. Metal Ceram.*, 1965, vol.4, pp.886-890.
43. B. Kieback and W. Schatt, "Anwendung Eines Kurzzeitigen Flussingphasensinterns fur die Herstellung von Fe-Ti-Sinterlegierungen," *Planseeber. Pulvermetall.*, 1980, vol.28, pp.204-215.
44. F. J. Puckert, W. A. Kaysser, and G. Petzow, "Transient Liquid Phase Sintering of Ni-Cu," *Z. Metallkde.*, 1983, vol.74, pp.737-743.
45. J. Puckert, W. A. Kaysser, and G. Petzow, " Dimensional Changes During Transient Liquid Phase Sintering of Fe-Ni," *Inter. J. Powder Met. Powder Tech.*, 1984, vol.20, pp.301-310.
46. L. I. Kivalo, V. V. Skorokhod, and V. Y. Petrishchev, "Dilatometric Investigation of the Sintering of Compacts from Titanium and Iron Powder Mixtures," *Soviet Powder Met. Metal Ceram.*, 1982, vol.21, pp.534-536.
47. L. I. Kivalo, V. V. Skorokhod, and N. F. Grigorenko, "Volume Changes Accompanying the Sintering of Compacts from Mixtures of Titanium and Iron Powders," *Soviet Powder Met. Metal Ceram.*, 1982, vol.21, pp.360-364.
48. L. I. Kivalo, V. V. Skorokhod, and N. F. Grigorenko, "Effect of Nickel on Sintering Processes in the Ti-Fe System. Part I," *Soviet Powder Met. Metal Ceram.*, 1983, vol.22, pp.543-546.
49. L. I. Kivalo, V. V. Skorokhod, and V. Y. Petrischchev, "Effect of Nickel on Sintering Processes in the Ti-Fe System. II. Dilatometric and Thermographic Investigation on the Sintering Process," *Soviet Powder Met. Metal Ceram.*, 1983, vol.22, pp.617-619.
50. V. V. Rana, S. M. Copley, and J. M. Whelan, "Sintering of Powder

Compacts with a Volatile Liquid Phase (MgO-LiF)," *Ceramic Microstructures '76*, R. M. Fulrath and J. A. Pask (eds.), Westview Press, Boulder, CO, 1977, pp.434-443.
51. E. M. Daver and W. J. Ullrich, "Hot Stage Microscopy Study of Liquid Phase Sintering," *Advanced Experimental Techniques in Powder Metallurgy*, J. S. Hirschhorn and K. H. Roll (eds.), Plenum Press, New York, NY, 1970, pp.189-200.
52. T. T. Lam, "High Density Sintering of Iron-Carbon Alloys Via Transient Liquid Phase," Report No. LBL-8001, Lawrence Berkeley Laboratory, University of California, Berkeley, CA, June 1978.
53. W. H. Rhodes, "Controlled Transient Solid Second-Phase Sintering of Yttria," *J. Amer. Ceramic Soc.*, 1981, vol.64, pp.13-19.
54. W. Kehl and H. F. Fischmeister, "Liquid Phase Sintering of Al-Cu Compacts," *Powder Met.*, 1980, vol.23, pp.113-119.
55. W. H. Baek, "Development of Transient Liquid Phase Sintering of Iron-Titanium," Ph.D. Thesis, Rensselaer Polytechnic Institute, Troy, NY, 1985.
56. S. Banerjee, V. Gemenetzis, and F. Thummler, "Liquid Phase Formation During Sintering of Low-Alloy Steels with Carbide-Base Master Alloy Additions," *Powder Met.*, 1980, vol.23, pp.126-129.
57. W. Schatt and H. J. Ullrich, "Metallographic-Electron Beam Microanalytical Investigations into the Liquid Phase Sintering of Cu-Ti," *Pract. Metallog.*, 1978, vol.15, pp.234-243.
58. C. Lequang, D. Treheux, J. Blanc-Benon, and P. Guiraldenq, "Metallic Powder-Liquid Metal Reaction Study: Correlation Between the Hardness Evolution and the Different States of the Ag-Sn-Hg System Sintering," *Sintering - New Developments*, M. M. Ristic (ed.), Elsevier Scientific, Amsterdam, Netherlands, 1979, pp.279-284.
59. M. Paulus, F. Laher-Lacour, P. Dugleux, and A. Dubon, "Defects and Transitory Liquid Phase Formation During the Sintering of Mixed Powders," *Trans. J. British Ceram. Soc.*, 1983, vol.82, pp.90-98.
60. M. R. Pickus, "Improving the Properties of P/M Steels Through Liquid Phase Sintering," *Inter. J. Powder Met. Powder Tech.*, 1984, vol.20, pp.311-323.
61. F. J. Puckert, W. A. Kaysser, and G. Petzow, "Densification and Pore Formation During Transient Liquid Phase Sintering of Ni-Cu," *Sci. Sintering*, 1984, vol.16, pp.105-113.
62. A. P. Savitskii and N. N. Burtsev, "Effect of Powder Particle Size on the Growth of Titanium Compacts During Liquid-Phase Sintering with Aluminum," *Soviet Powder Met. Metal Ceram.*, 1981, vol.20, pp.618-621.
63. A. P. Savitskii and N. N. Burtsev, "Compact Growth in Liquid Phase Sintering," *Soviet Powder Met. Metal Ceram.*, 1979, vol.18, pp.96-102.
64. S. J. Kiss, D. Cerovic, and E. Kostic, "Some Aspects of Transient Liquid Phase Sintering," *Sintering - Theory and Practice*, D. Kolar, S. Pejovnik and M. M. Ristic (eds.), Elsevier Scientific, Amsterdam, Netherlands, 1982, pp.251-256.
65. C. E. Bates and B. R. Patterson, "Transient Liquid Phase Sintering of P/M Titanium Alloys," Report 4805-XIV, Southern Research Institute, Birmingham, AL, February 1983.
66. J. Lorenz, J. Weiss, and G. Petzow, "Dense Silicon Nitride Alloys: Phase Relations and Consolidation, Microstructure and Properties," *Advances in Powder Technology*, G. Y. Chin (ed.), American Society for Metals, Metals Park, OH, 1982, pp.289-308.
67. G. Jangg, "Amalgams from the Point of View of Powder Metallurgy and

Sintering Technology," *Powder Met.*, 1964, vol.7, pp.241-250.
68. T. Kohno and M. J. Koczak, "Sintering and Dimensional Control of Mixed Elemental Bronze Powders," *Prog. Powder Met.*, 1982, vol.38, pp.463-481.
69. D. F. Berry, "Factors Affecting the Growth of 90/10 Copper/Tin Mixes Based on Atomized Powders," *Powder Met.*, 1972, vol.15, pp.247-266.
70. E. Peissker, "Pressing and Sintering Characteristics of Powder Mixtures for Sintered Bronze 90/10 Containing Different Amounts of Free Tin," *Modern Developments in Powder Metallurgy*, vol.7, H. H. Hausner and W. E. Smith (eds.), Metal Powder Industries Federation, Princeton, NJ, 1974, pp.597-614.
71. D. J. Lee and R. M. German, "Sintering Behavior of Iron-Aluminum Powder Mixes," *Inter. J. Powder Met. Powder Tech.*, 1985, vol.21, pp.9-21.
72. A. P. Savitskii, "Some Characteristic Features of the Sintering of Binary Systems," *Soviet Powder Met. Metal Ceram.*, 1980, vol.19, pp.488-493.
73. G. Langford and R. E. Cunningham, "Steel Casting by Diffusion Solidification," *Metall. Trans. B*, 1978, vol.9B, pp.5-19.
74. R. L. Coble, "Reactive Sintering," *Sintering - Theory and Practice*, D. Kolar, S. Pejovnik, and M. M. Ristic (eds.), Elsevier Scientific, Amsterdam, Netherlands, 1982, pp.145-151.
75. C. J. Quinn and D. L. Kohlstedt, "Reactive Processing of Titanium Carbide with Titanium, Part 1. Liquid-Phase Sintering," *J. Mater. Sci.*, 1984, vol.19, pp.1229-1241.
76. S. Boskovic, J. L. Gauckler, G. Petzow, and T. Y. Tien, "Reaction Sintering of Silica-Aluminum Nitride Mixture Forming Beta-Silicon Nitride Solid Solution," *Sintering - New Developments*, M. M. Ristic (ed.), Elsevier Scientific, Amsterdam, Netherlands, 1979, pp.374-380.
77. G. R. Terwilliger and F. F. Lange, "Pressureless Sintering of Si_3N_4," *J. Mater. Sci.*, 1975, vol.10, pp.1169-1174.
78. J. P. Hammond and G. M. Adamson, "Activated Sintering of Uranium Monocarbide," *Modern Developments in Powder Metallurgy*, vol.3, H. H. Hausner (ed.), Plenum Press, New York, NY, 1966, pp.3-25.
79. Y. Miyamoto, M. Koizumi, and O. Yamada, "High Pressure Self Combustion Sintering for Ceramics," *J. Amer. Ceramic Soc.*, 1984, vol.67, pp.C224-C225.
80. J. Mukerji, P. Greil, and G. Petzow, "Sintering of Silicon Nitride with a Nitrogen Rich Liquid Phase," *Sci. Sintering*, 1983, vol.15, pp.43-53.
81. O. Yamada, Y. Miyamoto, and M. Koizumi, "High Pressure Self-Combustion Sintering of Silicon Carbide," *Bull. Amer. Ceramic Soc.*, 1985, vol.64, pp.319-321.

CHAPTER EIGHT

Fabrication Concerns

A. Introduction

Throughout the prior presentation the emphasis has been on the fundamental mechanisms associated with liquid phase sintering. During the development of these fundamentals, the role of fabrication variables has been noted, but not emphasized. In this chapter the focus is orthogonal to that of the earlier chapters in that the independent, adjustable parameters are the primary concern. These adjustable parameters constitute the processing factors which have a significant effect on the success or failure of a liquid phase sintering cycle. Together the fundamental mechanisms and fabrication concerns constitute a matrix which determines the sintering behavior.

In industrial applications for liquid phase sintering dimensional control is a primary concern. The fabrication of precision and complex components requires control and uniformity in the dimensional changes. This is often a primary difficulty with materials processed by liquid phase sintering. Large dimensional changes are typical, often as large as 20% linear shrinkage. Because of these large shrinkages, supports for the compact are not practical except for simple shapes. During liquid phase sintering the compact is weak and friction with a support will cause uneven shrinkage and sometimes rupture. Furthermore, other common practical problems are slumping, distortion, and liquid migration due to gravitational forces (1-3). As a consequence, there are unique engineering design and processing concerns for liquid phase sintered compacts.

It is desirable to sinter to net shape for complex components. The topography of the specific component that can be sintered successfully depends on the amount of liquid phase formed and the rigidity of the solid skeleton. Typically, cantilevered sections will slump and large compacts will often rupture. As the component size is decreased, these problems become less critical, and control of the heating rate is often sufficient to control compact rigidity and insure uniform dimensions. Success in a production situation also depends on close control of the uniformity of the compact so that the dimensional changes are repeatable from compact to compact.

A common difficulty with liquid phase sintering is swelling during heating. For a given system the swelling tendency is established by basic material characteristics such as the diffusivity ratio (ratio of base diffusion in the additive to additive diffusion in the base) and solubility ratio (ratio of

TABLE 8.1

The Effect of Fabrication Variables on Liquid Phase Sintering

	Characteristic which favors	
Variable	Swelling	Shrinkage
solubility ratio	low	high
particle size	coarse	fine
green density	high	low
amount of liquid	small	large
contact angle	large	small
dihedral angle	small	large
time	short	long
temperature	low	high
heating rate	slow	rapid
additive homogeneity	low	high
powder porosity	low	high
atmosphere	insoluble	soluble

base solubility in the additive to additive solubility in the base). In spite of a swelling tendency, it is possible to select appropriate processing conditions to minimize swelling. As noted in this chapter, several decisions can be made to maximize densification. Table 8.1 gives a summary of these factors. For example, a large particle size favors swelling while a small particle size provides more densification. In systems like Cu-Sn, the strong commercial interest in the sintering characteristics has resulted in a full understanding of the role of the fabrication parameters (4,5). However, in other systems there is less information available. This summary of fabrication concerns provides a basis for initial optimization of liquid phase sintering cycles. Caution is necessary in using Table 8.1 because it is a simplified view of a complex process.

B. Particle Size

Particle size has a mixed effect on liquid phase sintering. Usually, the densification rate improves with a smaller particle size, giving higher final densities for a fixed processing cycle (6). A typical example of the particle size effect on sintering densification is shown in Figure 8.1 for TiC-Ni sintered at 1460°C for 2 hours (7). In the rearrangement stage, a small particle size improves the rate of rearrangement because of a large capillary force even though the amount of interparticle friction is increased. Likewise, in the solution-reprecipitation stage a small particle size improves the

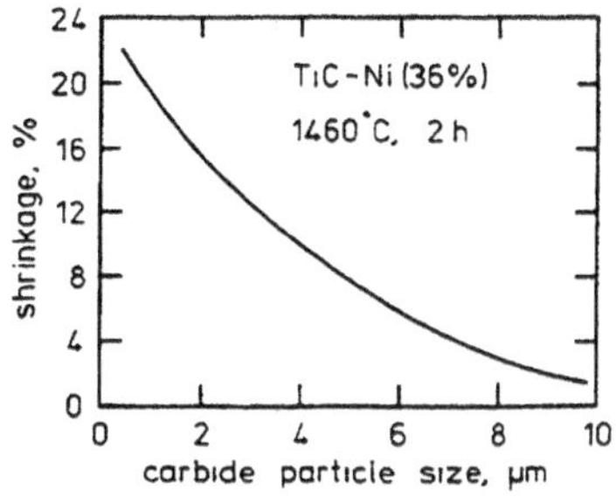

Figure 8.1 The shrinkage of TiC-36% Ni compacts sintered at 1460°C for 2 hours with various carbide particle sizes (7).

densification rate. It is a general result from studies involving persistent liquids that greater densification occurs at finer particle sizes (7-13). Also, the final product mechanical properties are improved by use of a small particle size (14). However, because many commercial processing cycles are relatively long, the benefits of a fine particle size can be lost because of microstructural coarsening.

A narrow particle size distribution aids densification. During the final stage of sintering large grains will induce exaggerated grain growth (15). Abnormally large grains can be nucleated by large particles in the initial compact. Also, the additive particle size will control the size of the pore formed at the prior additive particle sites; thus, small, uniform sized additive particles are desirable. For swelling systems, best densification is observed with small particle sizes and rapid heating rates to minimize swelling (16-18).

C. Particle Shape

The main effect of the particle shape is seen in compaction and in the initial rearrangement stage of liquid phase sintering. Spherical powders are undesirable in many situations because of a low green strength. Alternatively, there is difficulty in compacting fine, irregular powders because of the large interparticle friction. Thus, lower initial densities result from irregular powders; often this gives a lower sintered density. During rearrangement the capillary force varies with the particle shape. Generally, spherical particle shapes are more responsive to capillary action during rearrangement (19). The particle shape is also important to the uniformity of the sintered product. There is a greater chance of a nonuniform sintered microstructure with irregular particles than with spherical particles. In turn, such nonuniformity will degrade the properties (20). In the latter stages of liquid phase sintering grain shape changes due to solution-reprecipitation will eliminate the effects of the initial particle shape.

D. Internal Powder Porosity

Internal pores in the solid particles will preferentially fill with liquid and reduce the amount of intergranular liquid. The pores within the solid particles are typically smaller than the intergranular spaces. Thus, the capillary attraction of the liquid is much larger for the internal pores. As a consequence there is less liquid penetration between grains and differences in the sintering rate are observed as the particles become more porous.

The Fe-Cu alloys exhibit substantial differences in sintering behavior dependent on the Fe powder porosity (21-24). Figure 8.2 shows an example of the differences in shrinkage/swelling for Fe-Cu-C compacts with changes in the internal porosity of the Fe. The surface area increases as the internal porosity in the powder increases. The dense particles exhibit swelling due to liquid penetration between the solid grains, while the porous particles attract the liquid to the internal pores which reduces penetration and swelling. The effect of a higher carbon content is to reduce the grain boundary penetration by molten copper. Thus, higher carbon levels increase densification. A high compaction pressure can seal the internal pores and eliminate any effects from powder porosity.

E. Stoichiometry of the Powder

In sintering complex compounds the role of the powder stoichiometry is expected to be important because of the recognized composition effects on defect concentrations (25). However, in liquid phase sintering the role of the powder stoichiometry is poorly investigated. It appears that composition has a large effect on the liquid phase sintering of compounds. The effect is often related to the atomic defect structure (15,26-28). As an example, in the cemented carbides such as WC-Co, discontinuous grain growth is related to the stoichiometry of the carbide.

F. Additive Homogeneity

In persistent liquid phase sintering, the more homogeneous the formation of the liquid, the greater the densification. This has been clearly demonstrated in several experiments involving a range of materials (29-37). Additionally, a homogeneous powder mixture aids rapid alloying during liquid phase sintering and gives improved sintered properties. The distribution of the additive improves as the additive particle size decreases; thus, milling of the powder mixture prior to compaction is usually beneficial. Consequently, small additive particle sizes are most successful. However, homogenization of the mixture by annealing in the solid state proves detrimental to densification when the melt forms.

In a system exhibiting swelling during heating, minimal swelling occurs with a coated powder, where the additive is present as a uniform surface coating (30,38,39). The coating minimizes the size of the pores formed during interdiffusion prior to liquid formation and ensures a uniform distribution of liquid. Also, the amount of swelling can be reduced by use of a prealloyed additive to counteract a low solubility ratio (40,41). Recall from Chapter 4 the solubility ratio is defined as the solubility of the base in the additive divided by the additive solubility in the base.

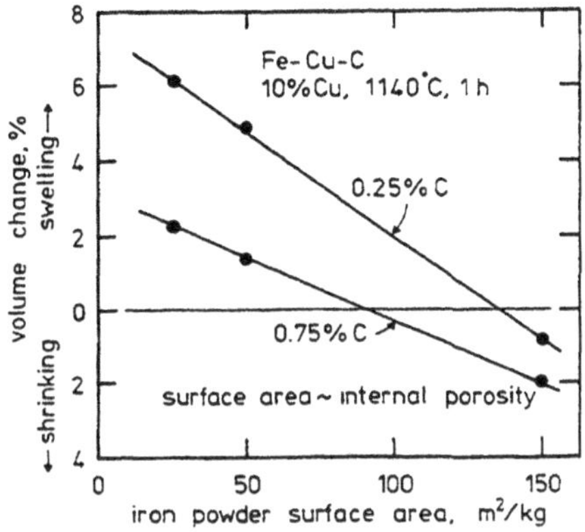

Figure 8.2 For Fe-Cu-C alloys swelling is decreased by carbon additions and a high level of internal porosity as measured by the iron powder surface area (23).

Clustering during the initial stage is the direct consequence of poor liquid distribution when the first melt forms. The need for uniformity in the sintering microstructure extends to particle packing. A uniform pore size is associated with a uniform particle packing. Uniform sized pores eliminate capillary flow to small pores, and thereby contribute to uniform sintering.

G. Amount of Additive

The amount of additive directly influences the volume fraction of liquid. Dimensional control and sintering kinetics are both dependent on the liquid content. As demonstrated earlier, the volume fraction of liquid has a significant influence on the sintering rate and the sintered microstructure. Features like the grain size, separation between grains, contiguity, grain shape accommodation, and liquid connectivity depend on the volume fraction of liquid. In the initial stage, the rearrangement force depends on the volume fraction of liquid. In the intermediate stage, densification and grain growth depend on the amount of liquid. Figure 8.3 shows the general effect of the volume fraction of liquid on the mechanisms operative in attaining specific densities. At high volume fractions of liquid, full density can be attained by solubility and rearrangement events alone. As the amount of additive decreases, the corresponding decrease in the volume fraction of liquid necessitates the action of solution-reprecipitation and final stage densification events to attain full density. As a consequence, the amount of additive is an important processing parameters which influences the rate of sintering, final microstructure, and sintered properties.

In a swelling system, an increase in the amount of additive affects the swelling. Figure 8.4 shows the effect of amount of liquid copper on the swelling of tungsten spheres when the contact angle is large (42). The compact of spherical tungsten powders was heated to 1100°C for 4 minutes. As shown in Chapter 4, greater initial stage swelling is observed as the

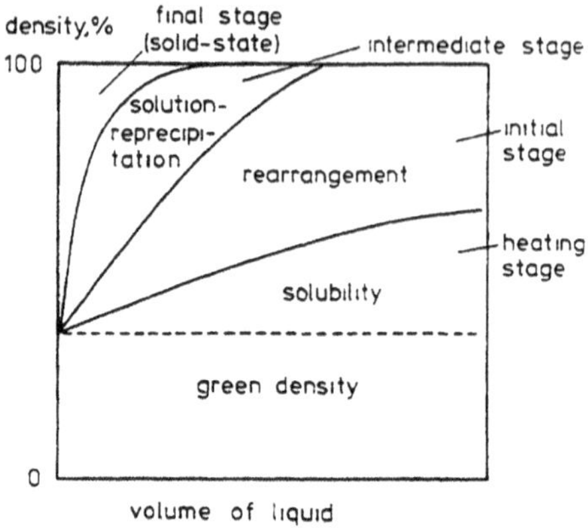

Figure 8.3 The stages necessary to attain various density levels during liquid phase sintering shown as functions of the liquid content.

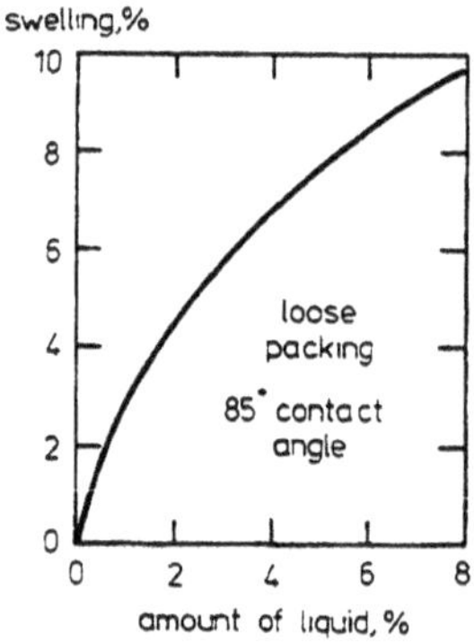

Figure 8.4 Swelling of loose packed tungsten particles versus the amount of copper for a high contact angle at 1100°C (42).

amount of liquid increases (43,44). Alternatively, in a high solubility ratio system densification is improved by higher liquid contents as demonstrated in Figure 8.5 for TiC-Ni (6). This figure shows the densification as a function of the sintering temperature for four different concentrations of nickel. In this case higher sintering temperatures are required to obtain equivalent degrees of densification as the volume of nickel is reduced. That is, the ease of attaining full density increases with the amount of liquid.

In the initial stage an increase in the liquid content results in a higher density after liquid flow. Although the capillary force decreases with an

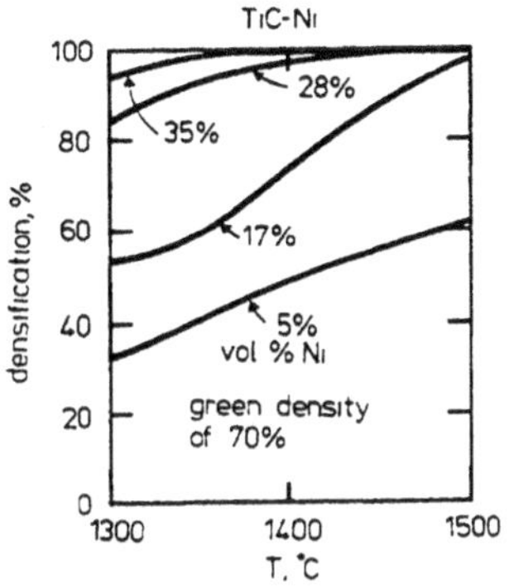

Figure 8.5 Densification of TiC-Ni compacts versus the sintering temperature, showing the easier densification at higher temperatures and higher liquid contents (6).

increase in liquid volume, the concurrent decrease in compact viscosity is more significant. With large quantities of liquid, complete densification can be obtained by rearrangement processes alone. However, dimensional control is lost and compact slumping is common for the high volume fractions of liquid. Thus, there is an optimal amount of liquid for initial densification (6).

In the intermediate stage, densification improves as the amount of liquid increases. Grain shape accommodation is necessary at low volume fractions of liquid. The greater the volume of liquid, the less shape accommodation and the greater the diffusive flux available for densification. Hence, faster intermediate stage densification is observed with high volume fractions of liquid.

In the final stage of liquid phase sintering, the volume of liquid has a major effect on the microstructure and the rate of microstructural coarsening. (6,45). Generally, large amounts of liquid are detrimental because of practical problems, while small amounts of liquid make full density difficult. As a compromise, 15 to 20% liquid is typical. Depending on the solid and liquid solubilities, the amount of additive is adjusted to provide approximately this quantity of liquid in several applications for liquid phase sintering.

H. Green Density

A high green density locks the microstructure and inhibits rearrangement (46). For a system with no solubility between the liquid and solid, rearrangement shrinkage decreases as the green density increases (11). A system with a high solubility ratio gives less shrinkage, but the final density increases directly with the green density. This is favorable for industrial applications of liquid phase sintering where minimal compact shape change is desired. Alternatively, a low solubility ratio system gives more swelling as the green density increases. The final porosity for a system showing expansion increases in proportion to the initial porosity. This is demonstrated for the Ti-Al system in Figure 8.6 (43). For an alloy with 30 at.% Al, the final porosity is shown as a function of the initial porosity. The behavior of

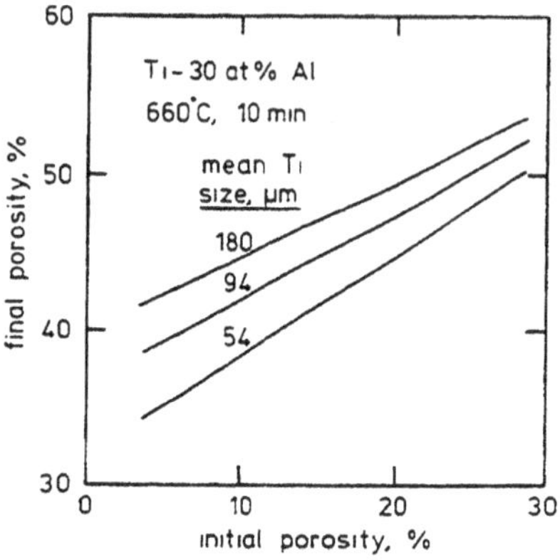

Figure 8.6 Porosity after sintering versus the initial porosity for Ti-Al compacts formed from three different titanium particle sizes (43).

three different titanium particle sizes are also shown in Figure 8.6; these data demonstrate increased swelling as the particle size increases. Although the final density improves with an increasing green density, the amount of swelling also increases (more dimensional change). However, the final density is still improved by a high green density. Thus, the benefits of a high green density are a higher sintered density and more compact rigidity.

Examples of the green density effect on sintered density are given in Figures 8.7 through 8.9. In Figure 8.7, the dimensional change on sintering Fe-Cu compacts is shown versus the density (47). At lower densities less swelling is obtained after sintering for 30 minutes at 1120°C. As the green density increases, there is a slight increase in the swelling. Dimensional growth can be minimized by adding carbon to these alloys or by using a slow heating rate. Heating rate is a factor because diffusional homogenization affects the amount of liquid formed. This behavior represents the combined roles of the solubility, melt penetration, and densification induced by the liquid phase. Swelling occurs during heating, and upon formation of the first melt further swelling is obtained because of penetration of interparticle contact points. This latter swelling event increases as the green density increases. Finally, at the sintering temperature, densification occurs as expected with a wetting liquid. At high green densities, the swelling which occurs during heating dominates the final density. Alternatively, as shown in Figure 8.8 for Al-Cu compacts, the sintered density is also a function of the green density for a high solubility ratio system (48). In this case the amount of liquid is relatively small. Because of the small amount of liquid, there is a strong sensitivity to the green density. For reference, the compaction pressure associated with each point is noted on Figure 8.8; the initial porosity decreases with a higher compaction pressure. A lower sintered porosity (higher density) is obtained with a higher green density. Green density is expected to have less of a role on the sintered density as the volume fraction of liquid is increased. This effect is shown for a tungsten

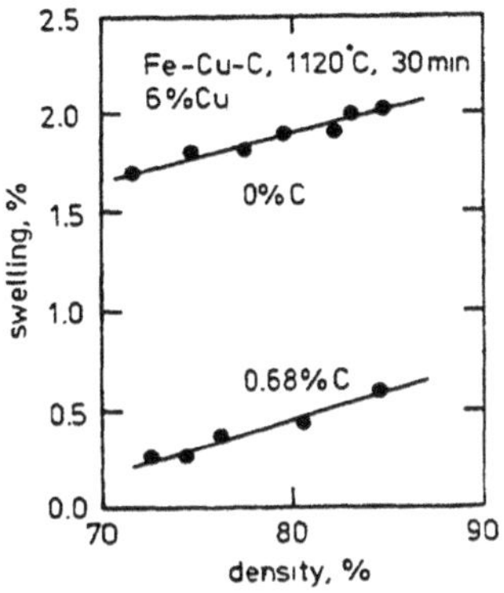

Figure 8.7 Swelling increases in Fe-Cu-C compacts as the density increases and as the carbon content decreases (47).

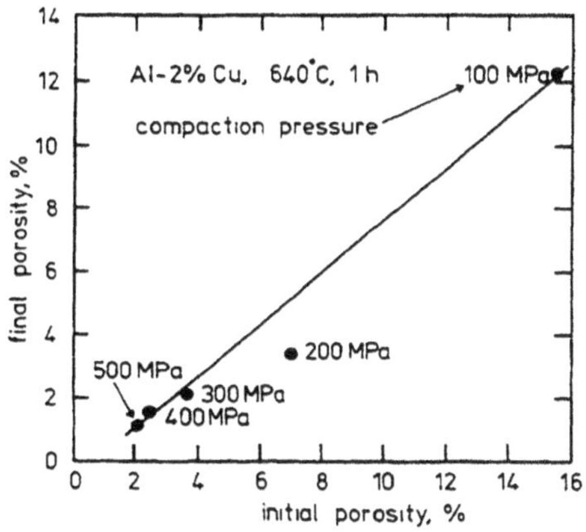

Figure 8.8 The sintered porosity shown as a function of the initial porosity for Al-2% Cu compacts sintered at 640°C for 1 hour (49).

heavy alloy in Figure 8.9, where slightly improved sintered densities are obtained with higher compaction pressures (green densities) (12). In this case the green density effect is not large because of the favorable solubility ratio and relatively large quantity of liquid. Indeed considerable densification occurs prior to liquid formation for the high solubility ratio systems.

The green density plays a role in determining the final density. Dimensional change decreases as the green density is increased for high solubility ratio systems (12,13,49-51). Because the dimensional change during

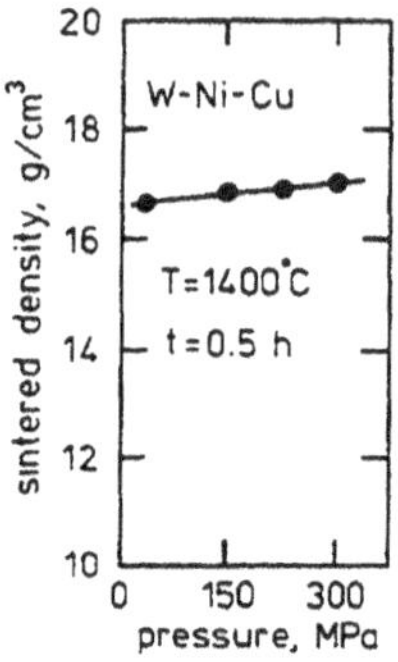

Figure 8.9 The compaction pressure effect on sintered density for a 93% W-5% Ni-2% Cu heavy alloy sintered for 30 minutes at 1400°C (12).

sintering scales with the green density, it is important to have a uniform green density to prevent uneven shrinkage. Alternatively, for low solubility ratio systems, a high green density increases the final density, but also increases the amount of dimensional growth. The densification obtained during the initial stage decreases as the green density (compaction pressure) is increased. Thus, intermediate stage densification becomes more important at the higher green densities.

I. Heating and Cooling Rates

Diffusional homogenization during heating can affect the amount of liquid formed at the sintering temperature. Additionally, homogenization will alter the initial liquid composition and thereby affect the dihedral and wetting angles. Furthermore, the heating rate has a critical effect on transient liquid phase systems as noted in Figure 7.15. Very slow heating can totally suppress the formation of a liquid, thus rapid heating has been found to be optimal in certain systems (52,53). Additionally, isothermal holds below the liquid formation temperature generally prove detrimental. In general, faster heating is beneficial, but is constrained by practical processing limitations such as oxide reduction, uniform heat transfer, lubricant burn-off, and binder removal.

During cooling from the sintering temperature, solid reprecipitation occurs due to a decreasing solid solubility in the liquid. This offers an opportunity to control the final mechanical properties through post-sintering heat treatments. The amount of reprecipitated solid can alter the strength (54-56). The greater the cooling rate, the more saturated the matrix will be with alloying additions; thus, higher sintered strengths are observed. At the solidification temperature for the liquid, rapid cooling can be harmful due to the formation of solidification porosity (54). These pores weaken a liquid phase sintered compact. In practice, cooling rate is not optimized; however, there are benefits possible from cooling rate control.

Another aspect of cooling rate is in controlling impurity segregation. As noted in Chapter 3, impurities which lower an interfacial energy will selectively segregate to the interface. The segregated species often result in embrittlement (57-60). For this reason rapid cooling is often beneficial since it prevents impurity segregation and the corresponding embrittlement.

J. Impurities and Trace Additives

Low concentrations of additives, both intentional and unintentional, can affect the basic system thermodynamics and kinetics during liquid phase sintering. Various roles of trace additives and impurities are possible. A common positive role is in controlling grain growth during the latter stages of densification (55,61-64). Another effect is to extend the solid solubility in the liquid and thereby increase the densification rate (65). It is common for the impurities to affect the wetting characteristics (66). The increased wetting is often coupled with extended solubility. For example, in W-Cu electrical contacts, Co, Ni, or P are added in small concentrations to improve wetting and solubility (67). Additionally, trace additives can be used to increase the amount of liquid (68,69) or to break down films which inhibit liquid formation (70). In ceramics, the impurities can form a grain boundary glassy phase which is liquid during sintering (71). This often leads to unintentional liquid phase sintering.

All of the above examples are beneficial effects of trace additives or impurities. Equally important are those which interfere with liquid phase sintering by decreasing the wetting, inducing rapid grain growth, or producing expansion (46,62,72). In this respect oxide control is quite significant to the liquid phase sintering of most metals and carbides. A good example of the oxide role is provided by the tool steels processed by supersolidus sintering. The initial powder can have up to 0.2% oxygen as a contaminant. To maintain control over the carbide content, excess carbon is required to reduce the oxygen contamination before liquid formation. Otherwise, the oxygen inhibits wetting and degrades the strength of the sintered product. Additionally, dissolved oxygen can lead to the formation of water bubbles inside a metal sintered in a reducing atmosphere (51). Hydrogen has a high solubility in many of the metals; however, water is relatively insoluble. The hydrogen reacts with the dissolved oxygen to form water vapor which subsequently nucleates pores, giving swelling and the formation of surface blisters. Oxygen can also form volatile oxides with various constituents and thereby alter the amount of liquid (72,73). Unfortunately, the largest concentrations of oxygen and other impurities are found with the smaller particle sizes which typically are sought for the better sintering densification (74).

K. Temperature

The primary requirement with respect to temperature is the formation of a liquid. Temperatures over the liquid formation temperature increase the diffusion rates, increase wetting, increase solubility of the solid in the liquid, decrease the liquid viscosity, and increase the amount of liquid. Also, there is less interfacial segregation at the higher temperatures. For these reasons there is a strong temperature sensitivity in persistent liquid phase sintering since all of these factors favor more rapid densification. However, there is a need to control the time-temperature combination to optimize densification and minimize microstructural coarsening. The optimal sintering time decreases as the sintering temperature is increased because of faster diffusion and a greater liquid content. A classic example of the temperature effect is provided by Price et al. (12) on tungsten heavy alloys. Figure 8.10 shows the sintered density dependence on sintering temperature for a W-Ni-Cu alloy. These data are for the same composition as shown in Figure 8.9. At the lower temperatures there is no liquid present, hence densification is totally dependent on solubility and diffusivity factors. At the higher temperatures, liquid forms and drastically increases the sintered density. However, at the higher temperatures there is more rapid grain growth and the compact tends to distort due to excess liquid formation.

In systems with a low solubility ratio, the final density is improved by a higher sintering temperature (44,75,76). Figure 8.11 shows the swelling for Al-Sn compacts versus the sintering temperature. This is a dilatometer trace showing swelling versus temperature during heating at a rate of 15°C/min. At the low temperatures swelling dominates the behavior. As the melting temperature of aluminum is approached, densification begins. Alternatively, a high solubility ratio system will exhibit densification prior to the formation of a liquid (31,77-79), as demonstrated in Figure 8.10 for W-Ni-Cu. Here the solubility and diffusivity variations with temperature are dominant.

L. Time

The time needed to attain full density depends on several processing factors, but is dominated by the volume fraction of solid and the sintering temperature (12,79). For a high solubility ratio system with over approximately 15% liquid, sintering times of 20 minutes are often satisfactory for full density. Figure 8.12 shows the effect of sintering time (on a logarithmic scale) on the sintered density of a W-Ni-Cu heavy alloy. These data are directly comparable with the green density and temperature data shown in Figures 8.9 and 8.10 for the same alloy. The major densification occurs in the first 20 minutes at temperature; beyond approximately 60 minutes there is little densification gain. For incompletely densified material extended times are beneficial to the sintered properties because of continued pore elimination. However, prolonged sintering is unfavorable because of pore growth, preferential vaporization, and microstructural coarsening (76,78). Overall, the densification equations show that time is a relatively weak processing factor in comparison with parameters like the sintering temperature. Accordingly, large changes in the sintering time are necessary to bring about any significant improvements in sintered properties.

M. Atmosphere

The sintering atmosphere is a final processing factor of significance to liquid phase sintering. The atmosphere protects against surface contamination during sintering. Additionally, a cleaning function is desired from the atmosphere to remove films (for example oxides) for rapid melt flow. In many systems best densification and properties have been achieved using vacuum sintering. Residual atmospheres can be trapped in pores which seal during liquid flow. Such trapped atmospheres inhibit full densification as noted in Chapter 6 (6,11,79,80). Sintering in an inert or insoluble atmosphere is most detrimental for this reason. Figure 8.13 shows the atmosphere effect on densification for W-Cu sintered in hydrogen and vacuum (11). In spite of the

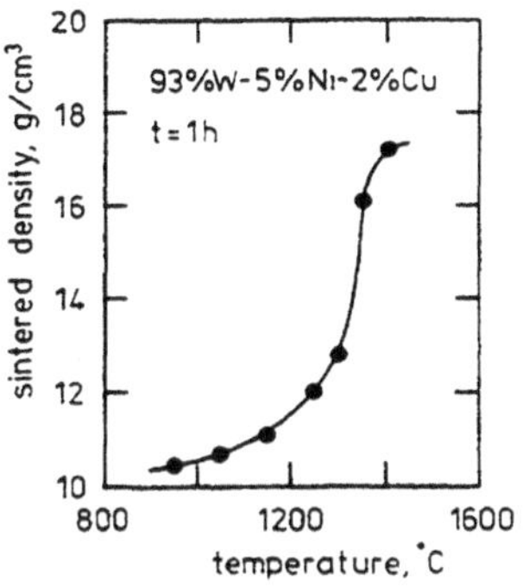

Figure 8.10 Sintering temperature effect on the sintered density of a W-Ni-Cu heavy alloy sintered for 1 hour (12). Liquid formation greatly aids densification.

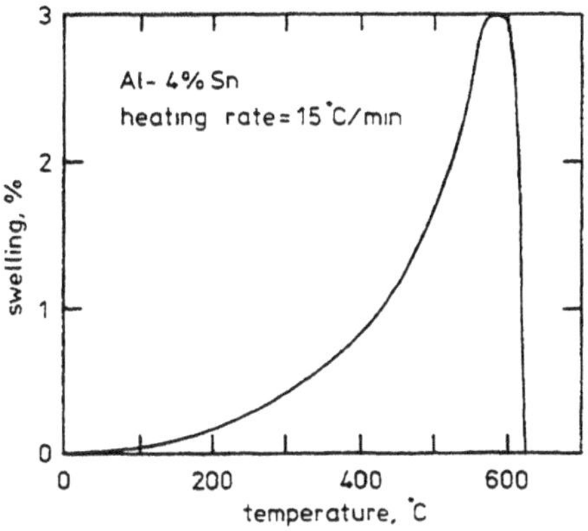

Figure 8.11 Swelling versus temperature for a Al-Sn compact during heating at a rate of 15°C/min, with eventual densification at the high temperatures (76).

solubility of hydrogen in copper, vacuum proved to be superior since there is no impediment to pore closure. Beyond pore stabilization, the atmosphere plays a role in reducing surface films which can alter the dihedral or wetting angles.

N. Summary

This chapter has addressed the various fabrication concerns associated with liquid phase sintering. The fabrication parameters are a network of independent and adjustable variables which can have a significant influence on

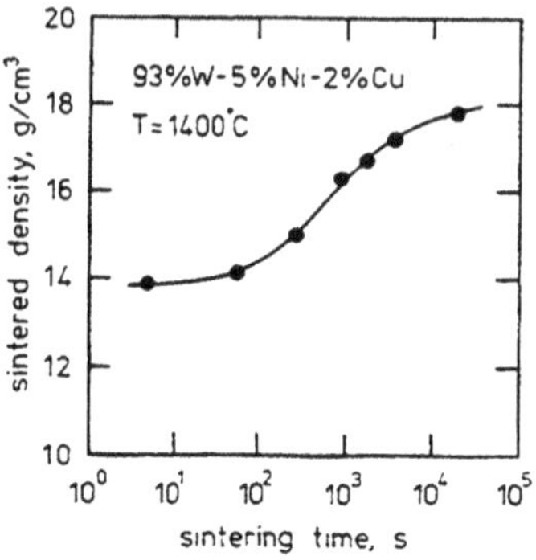

Figure 8.12 The effect of sintering time on the density of W-Ni-Cu heavy alloy sintered at 1400°C (12).

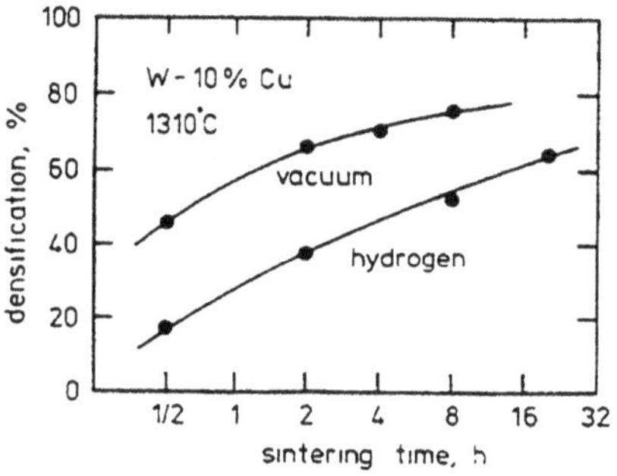

Figure 8.13 Densification versus sintering time for W-10% Cu sintered at 1310°C in either hydrogen or vacuum (11).

the rate of sintering, final density, and sintered microstructure. The successful application of liquid phase sintering theory to specific material systems is dependent on selection of the appropriate fabrication cycle for optimal properties. As demonstrated in this chapter, the definition of optimal processing depends on three considerations; i) the shape complexity of the compact, ii) the material characteristics (such as the solubility ratio), and iii) the fabrication variables. Temperature, particle size, and amount of additive are probably the most important fabrication variables. A high sintering temperature makes for more rapid sintering, but can lead to rapid microstructural coarsening. Small particle sizes are typically beneficial because of the greater capillary force and more homogeneous microstructure. However, small particles provide some practical difficulties of poor compactibility and high interparticle forces. Often, particles of 1 to 10 μm prove most suitable to liquid phase sintering treatments. Finally, the amount of liquid is controlled by the amount of additive. The liquid content affects the densification rate, stages of sintering, transport mechanisms, microstructure, and final compact properties. Optimization of liquid phase sintering processes requires that these three variables (temperature, particle size, and amount of additive) be carefully controlled to ensure repeatable results. Beyond these three, other characteristics (such as surface oxides, sintering time, sintering atmosphere, particle shape, green density, mix homogeneity, impurities, powder porosity, heating rate, and compound stoichiometry) can have a variable influence on the sintered product. This chapter has provided examples of the influences of these various fabrication factors, with an aim at understanding the more practical factors related to liquid phase sintering.

O. References

1. E. G. Zukas, P. S. Z. Rogers, and R. S. Rogers, "Unusual Spheroid Behavior During Liquid-Phase Sintering," *Inter J. Powder Met. Powder Tech.*, 1977, vol.13, pp.27-38.
2. A. N. Niemi and T. H. Courtney, "Settling in Solid-Liquid Systems with Specific Application to Liquid Phase Sintering," *Acta Met.*, 1983, vol.31, pp.1393-1401.
3. T. H. Courtney, "Densification and Structural Development in Liquid Phase Sintering," *Metall. Trans. A*, 1984, vol.15A, pp.1065-1074.
4. G. Dowson, "The Sintering of Bronze," *Metal Powder Rep.*, 1984, vol.39, pp.71-73.
5. D. F. Berry, "Factors Affecting the Growth of 90/10 Copper/Tin Mixes Based on Atomized Powders," *Powder Met.*, 1972, vol.15, pp.247-266.
6. V. N. Eremenko, Y. V. Naidich, and I. A. Lavrinenko, *Liquid Phase Sintering*, Consultants Bureau, New York, NY, 1970.
7. T. J. Whalen and M. Humenik, "Sintering in the Presence of a Liquid Phase," *Sintering and Related Phenomena*, G. C. Kuczynski, N. Hooton and C. Gibbon (eds.), Gordon and Breach, New York, NY, 1967, pp.715-742.
8. L. Froschauer and R. M. Fulrath, "Direct Observation of Liquid-Phase Sintering in the System Tungsten Carbide-Cobalt," *J. Mater. Sci.*, 1976, vol.11, pp.142-149.
9. Y. V. Naidich, I. A. Lavrinenko, and V. A. Evdokimov, "Densification During Liquid-Phase Sintering in Diamond-Metal Systems," *Soviet Powder Met. Metal Ceram.*, 1972, vol.11, pp.715-718.
10. L. Froschauer and R. M. Fulrath, "Direct Observation of Liquid-Phase Sintering in the System Iron-Copper," *J. Mater. Sci.*, 1975, vol.10, pp.2146-2155.
11. H. S. Cannon and F. V. Lenel, "Some Observations on the Mechanism of Liquid Phase Sintering," *Plansee Proceedings*, F. Benesovsky (ed.), Metallwerk Plansee, Reutte, Austria, 1953, pp.106-121.
12. G. H. S. Price, C. J. Smithells, and S. V. Williams, "Sintered Alloys. Part I - Copper-Nickel-Tungsten Alloys Sintered with a Liquid Phase Present," *J. Inst. Metals*, 1938, vol.62, pp.239-264.
13. F. V. Lenel, "Sintering with a Liquid Phase," *The Physics of Powder Metallurgy*, W. E. Kingston (ed.), McGraw-Hill, New York, NY, 1951, pp.238-253.
14. H. Suzuki and K. Hayashi, "The Role of Particle Size and Carbon Content in High Strength WC-10% Co Alloys," *Trans. Japan Inst. Metals*, 1969, vol.10, pp.360-364.
15. M. Schreiner, T. Schmitt, E. Lassner, and B. Lux, "On the Origins of

Discontinuous Grain Growth During Liquid Phase Sintering of WC-Co Cemented Carbides," *Powder Met. Inter.*, 1984, vol.16, pp.180-183.
16. T. E. Volin and R. W. Balluffi, "Annealing Kinetics of Voids and the Self-Diffusion Coefficient in Aluminum," *Phys. Stat. Solidi*, 1968, vol.25, pp.163-173.
17. L. I. Kivalo, V. V. Skorokhod, and V. Y. Petrischchev, "Effect of Nickel on Sintering Processes in the Ti-Fe System. II. Dilatometric and Thermographic Investigation on the Sintering Process," *Soviet Powder Met. Metal Ceram.*, 1983, vol.22, pp.617-619.
18. K. V. Savitskii, V. I. Itin, Y. I. Kozlov, and A. P. Savitskii, "The Effect of the Particle Size of Aluminum Powder on the Sintering of a Cu-Al Alloy in the Presence of a Liquid Phase," *Soviet Powder Met. Metal Ceram.*, 1965, vol.4, pp.886-890.
19. M. F. Yan, "Sintering of Ceramics and Metals," *Advances in Powder Technology*, G. Y. Chin (ed.), American Society for Metals, Metals Park, OH, 1982, pp.99-133.
20. J. N. Brecker, "Analysis of Bond Formation in Vitrified Abrasive Wheels," *J. Amer. Ceramic Soc.*, 1974, vol.57, pp.486-489.
21. F. V. Lenel and T. Pecanha, "Observations on the Sintering of Compacts from a Mixture of Iron and Copper Powders," *Powder Met.*, 1973, vol.16, pp.351-365.
22. K. Tabeshfar and G. A. Chadwick, "Dimensional Changes During Liquid Phase Sintering of Fe-Cu Compacts," *Powder Met.*, 1984, vol.27, pp.19-24.
23. S. J. Jamil and G. A. Chadwick, "Investigation and Analysis of Liquid Phase Sintering of Fe-Cu and Fe-Cu-C Compacts," *Proceedings Sintering Theory and Practice Conference*, The Metals Society, London, UK, 1984, pp.13.1-13.14.
24. K. Tabeshfar and G. A. Chadwick, "The Role of Powder Characteristics and Compacting Pressure in Liquid Phase Sintering of Fe-Cu Compacts," *Proceedings P/M-82*, Associazione Italiana di Metallurgia, Milano, Italy, 1982, pp.693-700.
25. H. U. Anderson and M. C. Proudian, "Sintering of Lithium Fluoride Fluxed Strontium Titanate," *Sintering and Heterogeneous Catalysis*, G. C. Kuczynski, A. E. Miller, and G. A. Sargent (eds.), Plenum Press, New York, NY, 1984, pp.281-292.
26. A. I. Kingon and J. B. Clark, "Sintering of PZT Ceramics; II, Effect of PbO Content on Densification," *J. Amer. Ceramic Soc.*, 1983, vol.66, pp.256-260.
27. H. E. Exner, E. Santa Marta, and G. Petzow, "Grain Growth in Liquid-Phase Sintering of Carbides," *Modern Developments in Powder Metallurgy*, vol.4, H. H. Hausner (ed.), Plenum Press, New York, NY, 1971, pp.315-325.
28. K. Y. Eun, "The Abnormal Grain Growth and the Effect of Ni Substitution on Mechanical Properties in Sintered WC-Co Alloys," Ph.D. Thesis, Korea Advanced Institute of Science and Technology, Seoul, Korea, 1983.
29. W. J. Huppmann, H. Riegger, W. A. Kaysser, V. Smolej, and S. Pejovnik, "The Elementary Mechanisms of Liquid Phase Sintering. I. Rearrangement," *Z. Metallkde.*, 1979, vol.70, pp.707-713.
30. P. Ramakrishnan and B. K. Agrawal, "Influence of Powder Characteristics on the Sintering of Aluminum-Tin Powder," *Inter. J. Powder Met.*, 1969, vol.5, no.2, pp.79-87.
31. R. F. Snowball and D. R. Milner, "Densification Processes in the

Tungsten Carbide-Cobalt System," *Powder Met.*, 1968, vol.11, pp.23-40.
32. M. Lejbrandt and W. Rutkowski, "The Effect of Grain Size of Nickel Activating the Sintering of Molybdenum," *Planseeber. Pulvermetall.*, 1977, vol.25, pp.3-12.
33. K. V. Sebastian and G. S. Tendolkar, "Densification in W-Cu Sintered Alloys Produced from Coreduced Powders," *Planseeber. Pulvermetall.*, 1977, vol.25, pp.84-100.
34. J. Kurtz, "Sintered High Density Tungsten and Tungsten Alloys," *Proceedings Second Annual Spring Meeting*, Metal Powder Association, New York, NY, 1946, pp.40-52.
35. J. C. Billington, C. Fletcher, and P. Smith, "Iron-Copper-Tin Sintered Compacts," *Powder Met.*, 1973, vol.16, pp.327-350.
36. R. H. Arendt, "Liquid-Phase Sintering of Magnetically Isotropic and Anisotropic Compacts of $BaFe_{12}O_{19}$ and $SrFe_{12}O_{19}$," *J. Appl. Phys.*, 1973, vol.44, pp.3300-3305.
37. W. J. Huppmann and W. Bauer, "Characterization of the Degree of Mixing in Liquid-Phase Sintering Experiments," *Powder Met.*, 1975, vol.18, pp.249-258.
38. I. M. Fedorchenko, I. I. Ivanova, and O. I. Fushchich, "Of Sintering Metal Powders on the Mechanism of Activating the Process," *Modern Developments in Powder Metallurgy*, vol.4, H. H. Hausner (ed.), Plenum Press, New York, NY, 1971, pp.255-266.
39. F. V. Lenel and K. S. Hwang, "The Mechanical Properties of High-Density Iron-Copper Alloys from a Composite Powder," *Powder Met. Inter.*, 1980, vol.12, pp.88-90.
40. P. Ramakrishnan, " The Effect of Powder Characteristics on the Sintering of a Multicomponent System," *J. Powder Bulk Solids Tech.*, 1979, vol.3, pp.32-34.
41. D. J. Lee and R. M. German, "Sintering Behavior of Iron-Aluminum Powder Mixes," *Inter. J. Powder Met. Powder Tech.*, 1985, vol.21, pp.9-21.
42. W. J. Huppmann and R. Riegger, "Modelling of Rearrangement Processes in Liquid Phase Sintering," *Acta Met.*, 1975, vol.23, pp.965-971.
43. A. P. Savitskii and N. N. Burtsev, "Effect of Powder Particle Size on the Growth of Titanium Compacts During Liquid-Phase Sintering with Aluminum," *Soviet Powder Met. Metal Ceram.*, 1981, vol.20, pp.618-621.
44. K. V. Savitskii, V. I. Itin, and Y. I. Kozlov, "The Mechanism of Sintering of Copper-Aluminum Powder Alloys in the Presence of a Liquid Phase," *Soviet Powder Met. Metal Ceram.*, 1966, vol.5, pp.4-9.
45. K. W. Lay, "Grain Growth in Urania-Alumina in the Presence of a Liquid Phase," *J. Amer. Ceramic Soc.*, 1968, vol.51, pp.373-376.
46. W. Kehl and H. F. Fischmeister, "Liquid Phase Sintering of Al-Cu Compacts," *Powder Met.*, 1980, vol.23, pp.113-119.
47. T. Krantz, "Effect of Density and Composition on the Dimensional Stability and Strength of Iron-Copper Alloys," *Inter. J. Powder Met.*, 1969, vol.5, no.3, pp.35-43.
48. Y. E. Geguzin and Y. I. Klinchuk, "Mechanisms and Kinetics of the Initial Stage of Solid-Phase Sintering of Compacts from Powders of Crystalline Solids," *Soviet Powder Met. Metal Ceram.*, 1976, vol.15, pp.512-518.
49. L. S. Martsunova, A. P. Savitskii, E. N. Ushakova, and B. I. Matveev, "Sintering of Aluminum with Copper Additions," *Soviet Powder Met. Metal Ceram.*, 1973, vol.12, pp.956-959.
50. I. H. Moon and Y. S. Kwon, "The Relationship Between Compactibility

and Shrinkage Rate in Activated Sintering of Nickel-Doped Tungsten," *Powder Met.*, 1974, vol.17, pp.363-369.
51. V. A. Dymchenko and A. P. Popovich, "Hydrogen Sickness of Sintered Copper," *Soviet Powder Met. Metal Ceram.*, 1983, vol.22, pp.347-349.
52. C. Guyard, C. H. Allibert, J. Driole, and G. Raisson, "Liquid Phase Sintering of Prealloyed Powders of Co-Base Alloy," *Sci. Sintering*, 1981, vol.13, pp.149-163.
53. F. J. Puckert, W. A. Kaysser, and G. Petzow, "Transient Liquid Phase Sintering of Ni-Cu," *Z. Metallkde.*, 1983, vol.74, pp.737-743.
54. T. K. Kang, E. T. Henig, W. A. Kaysser, and G. Petzow, "Effect of Cooling Rate on the Microstructure of a 90W-7Ni-3Fe Heavy Alloy," *Modern Developments in Powder Metallurgy*, vol.14, H. H. Hausner, H. W. Antes and G. D. Smith (eds.), Metal Powder Industries Federation, Princeton, NJ, 1981, pp.189-203.
55. B. Lux, G. Jangg, and H. Danninger, "The Influence of Chemistry and Various Fabrication Parameters on the Properties of Tungsten Heavy Metals," Final Report DAER078-G-086, Institut fur Chemische Technologie Anorganischer Stoffe, Technical Universtitat Wien, Vienna, Austria, February 1981.
56. S. S. Kang and D. N. Yoon, "The Effect of Cooling Rate on the Strength of Sintered Fe-Cu Compacts," *Powder Met.*, 1977, vol.20, pp.70-73.
57. B. C. Muddle and D. V. Edmonds, "Interfacial Segregation and Embrittlement in Liquid Phase Sintered Tungsten Alloys," *Metal Sci.*, 1983, vol.17, pp.209-218.
58. H. Blumenthal and R. Silverman, "A Study of the Microstructure of Titanium Carbide," *Trans. AIME*, 1955, vol.203, pp.317-322.
59. C. Lea, B. C. Muddle, and D. V. Edmonds, "Segregation to Interphase Boundaries in Liquid-Phase Sintered Tungsten Alloys," *Metall. Trans. A*, 1983, vol.14A, pp.667-677.
60. M. R. Eisenmann and R. M. German, "Factors Influencing Ductility and Fracture Strength in Tungsten Heavy Alloys," *Inter. J. Refractory Hard Met.*, 1984, vol.3, pp.86-91.
61. J. White, "Microstructure and Grain Growth in Ceramics in the Presence of a Liquid Phase," *Sintering and Related Phenomena*, G. C. Kuczynski (ed.), Plenum Press, New York, NY, 1973, pp.81-108.
62. E. Lassner, M. Schreiner, and B. Lux, "Influence of Trace Elements in Cemented Carbide Production: Part 2," *Inter. J. Refractory Hard Metals*, 1982, vol.1, pp.97-102.
63. J. P. Sadocha and H. W. Kerr, "Grain-Ripening of the Lead-Rich Phase in Partially Molten Pb-Sb-Sn and Pb-Sb-Sn-As Solders," *Metal Sci. J.*, 1973, vol.7, pp.138-146.
64. Y. Bin and F. V. Lenel, "Activated Sintering of Molybdenum Powder Electroless Plated with a Nickel-Phosphorus Alloy," *Inter. J. Powder Met. Powder Tech.*, 1984, vol.20, pp.15-21.
65. V. I. Itin, K. V. Savitskii, Y. I. Kozlov, and A. D. Bratchikov, "Activation of Densification in the Sintering of Some Powder Metallurgy Alloys. 1. Influence of Silver on the Shrinkage Kinetics of Cu-Al Alloys," *Soviet Powder Met. Metal Ceram.*, 1969, vol.8, pp.635-641.
66. W. H. Lenz and J. M. Taub, "Liquid Oxide Phase Sintering of Molybdenum and Tungsten," *J. Less-Common Metals*, 1961, vol.3, pp.429-432.
67. I. H. Moon and J. S. Lee, "Activated Sintering of Tungsten-Copper Contact Materials," *Powder Met.*, 1979, vol.22, pp.5-7.

68. J. F. Shackelford, P. S. Nicholson, and W. W. Smeltzer, "Influence of Silica on Sintering of Partially Stabilized Zirconia," *Bull. Amer. Ceramic Soc.*, 1974, vol.53, pp.865-867.
69. J. Sautereau and A. Mocellin, "Sintering Behaviour of Ultrafine NbC and TaC Powders," *J. Mater. Sci.*, 1974, vol.9, pp.761-771.
70. J. W. Butcher and J. N. Lowe, "Activated Sintering in Beryllium Powders by Selective Addition of Trace Elements," *Beryllium Technology*, vol.1, L. M. Schetky and H. A. Johnson (eds.), Gordon and Breach, New York, NY, 1966, pp.501-522.
71. P. E. D. Morgan and M. S. Koutsoutis, "Phase Studies Concerning Sintering in Aluminas Doped with Ti(+4)," *J. Amer. Ceramic Soc.*, 1985, vol.68, pp.C156-C158.
72. R. Wahling, P. Beiss, and W. J. Huppmann, "Sintering Behaviour and Performance Data of HSS-Components," *Proceedings Sintering Theory and Practice Conference*, The Metals Society, London, UK, 1984, pp.15.1-15.5.
73. F. Frehn and W. Hotop, "Effect of Small Boron Contents on the Properties of Compacts Prepared by Vacuum Sintering," *Symposium on Powder Metallurgy*, Special Report 58, Iron and Steel Institute, London, UK, 1956, pp.137-143.
74. S. K. Chatterjee, J. V. Castell-Evans, and P. A. Ainsworth, "Influence of Tin Powder Characteristics on the Sintering of Iron-Tin-Copper Powder Compacts," *Powder Met.*, 1972, vol.15, pp.153-165.
75. N. C. Kothari and J. Waring, "Sintering Kinetics in Iron-Copper Alloys with and without a Liquid Phase," *Powder Met.*, 1964, vol.7, pp.13-33.
76. R. Sundaresan and P. Ramakrishnan, "Liquid Phase Sintering of Aluminum Base Alloys," *Inter. J. Powder Met. Powder Tech.*, 1978, Vol.14, pp.9-16.
77. I. H. Moon and Y. S. Kwon, "Some Observations on Sintering of the Nickel-Doped Tungsten Compacts," *Scripta Met.*, 1979, vol.13, pp.33-36.
78. B. Meredith and D. R. Milner, "The Liquid-Phase Sintering of Titanium Carbide," *Powder Met.*, 1976, vol.19, pp.162-170.
79. R. J. Nelson and D. R. Milner, "Densification Processes in the Tungsten Carbide-Cobalt System," *Powder Met.*, 1972, vol.15, pp.346-363.
80. R. M. German and K. S. Churn, "Sintering Atmosphere Effects on the Ductility of W-Ni-Fe Heavy Metals," *Metall. Trans. A*, 1984, vol.15A, pp.747-754.

CHAPTER NINE

Properties of Liquid Phase Sintered Materials

A. Typical Behavior

In this discussion, the concern is with the properties of systems with high volume fractions of solid (typically over 50% solid at the sintering temperature), which includes most practical materials processed by liquid phase sintering techniques. There is a diversity of final microstructures possible by liquid phase sintering. The microstructure carries over to affect the properties, especially mechanical behavior. However, microstructure is not the only factor affecting the properties of liquid phase sintered materials (1,2). Table 9.1 lists the specific factors by categories of powder characteristics, sintering cycle, alloy composition, post-sintering heat treatment, sintered microstructure, and testing conditions. In light of such diversity, it is difficult to make specific statements about optimal conditions. However, there are some general results which provide insight to the links between composition, processing, microstructure, and properties.

For the most part, liquid phase sintered materials consist of a hard grain dispersed in a matrix which was liquid at the sintering temperature. At this point it is appropriate to refer to the solidified liquid phase as the matrix, since it is no longer liquid. Often the matrix has a higher ductility and toughness than the solid grains. For the low dihedral angles (below 60°) typical to liquid phase sintering, the matrix phase is continuous as discussed in Chapter 2. Also the grain structure is interconnected for all dihedral angles over 0°. The hard grains provide strengthening while the matrix aids densification and increases the composite toughness. Mechanical properties are a dominant concern. Additionally, there is interest in the electrical, magnetic, wear, oxidation, creep, thermal, radiation, and dielectric properties of liquid phase sintered materials. Unfortunately, these latter concerns are not as well understood as the mechanical behavior (3). For this reason, the emphasis in this chapter will be on properties such as strength, ductility, elastic modulus, and fracture resistance.

B. Microstructure Effects on Mechanical Behavior

It is characteristic of liquid phase sintered materials that pore elimination is a primary concern in attaining maximum mechanical properties. For the typical structure of a hard grain dispersed in a softer matrix, the mechanical properties in the fully consolidated condition depend on the properties of the individual phases and their relation to each other as expressed by the

TABLE 9.1

Factors Affecting Mechanical Behavior

Powders	Sintering	Composition
size	time	solid content
shape	temperature	impurities
purity	atmosphere	binder
packing	heating rate	intermetallics
mixing	green density	
compressibility	thermal gradients	
	component size	

Microstructure	Testing	Heat-Treatment
grain size	surface finish	temperature
neck size	strain rate	time
contiguity	test temperature	atmosphere
grain separation	stress concentrations	cooling rate
porosity	specimen size	
pore size	stress state	
segregation	flaws	

microstructure (4). Fischmeister and Karlsson (5) provide a review of the several microstructure-property relations applicable to two phase composite materials, of which liquid phase sintered materials are a subset. In many instances, best overall properties can be expected using matrix layers of approximately 1 μm thickness. Alternatively, for a ductile grain and brittle matrix, very thin matrix layers are preferable. The combination of a brittle matrix and brittle grain generally produces a brittle composite, while the ductile matrix and grain combination will give a ductile composite. These generalizations assume homogeneous structures, lacking embrittling intermetallics, segregated impurities, residual pores, and thermal strains.

The properties associated with failure (like ductility, fracture strength, and fracture toughness) are most sensitive to microstructure variations. For example, ductility is very sensitive to changes in purity, grain size, grain separation, and intermetallic precipitates. Alternatively, the elastic modulus is typically sensitive only to the amount of solid phase and the contiguity. These various mechanical property sensitivities to the microstructure are developed in the balance of this section, starting with a discussion on hardness.

1. Hardness

Hardness has a fairly straightforward dependance on the microstructure. For the typical case of a harder solid grain dispersed in the solidified matrix, the matrix has the greater ductility. During deformation, the hard grains provide strengthening to the matrix and decrease the dislocation motion (assuming the matrix is crystalline). However, the matrix properties control the overall hardness (6). As a consequence the composite hardness depends on the volume fraction of the harder phase (7-10). The hardness decreases as the amount of matrix increases as illustrated in Figure 9.1a for various WC-Co alloys. Also, the hardness decreases as the grain size increases as shown in Figure 9.1a, due to a greater separation between grains (8,10,11). Accordingly, a small mean grain separation is beneficial as demonstrated in Figure 9.1b (7,8,10,12). Finally, a high contiguity aids the hardness because of greater rigidity from the solid-solid contacts.

Gurland (13) has suggested a hardness *H* dependence on the volume fraction of solid V_S and hardnesses of the matrix H_L and grains H_S as follows:

$$H = H_L (1 - C_{SS} V_S) + H_S C_{SS} V_S \tag{9.1}$$

with C_{ss} equal to the carbide contiguity.

For the cemented carbides it also has been demonstrated that the hardness varies with the inverse square root of the mean grain separation. Alternatively, Grathwohl and Warren (7) suggest an equation for the hardness dependence on microstructure which incorporates the solid skeleton contribution as well as that of each phase individually. Their model considers the hard grains as dispersion strengtheners as follows:

$$H = a H_S C_{SS} + b H_S V_S + H_L V_L (1 - G^{-1/2}) \tag{9.2}$$

where *G* is the solid grain size, and *a* and *b* are empirical constants. From these models it is evident that hardness is highest for a material with a high volume fraction of solid, fine grain size, and high contiguity.

2. Elastic Modulus

The elastic modulus is dependent on the amount of hard phase and the contiguity of the hard phase. Thus, the elastic modulus will be reduced by a high volume fraction of matrix phase (14,15). Several models are available linking the elastic modulus to the volume fraction of solid (4). For the case of isolated hard grains, Paul (16) assumes equivalent strains in the two phases to give an upper bound estimate on the composite elastic modulus *E* as follows:

$$E = E_S V_S + E_L V_L \tag{9.3}$$

where *V* represents the volume fraction, and the subscripts *S* and *L* represent the solid and matrix (liquid) phases, respectively. As the connectivity

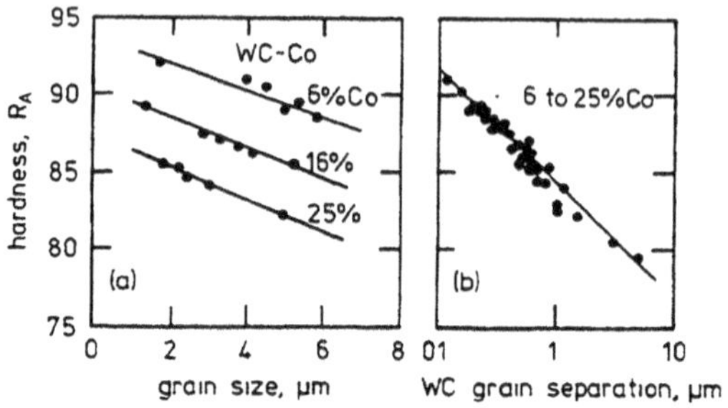

Figure 9.1 The effect of grain size (a) and grain separation (b) on the hardness of WC-Co alloys (8).

of the microstructure increases, a lower bound composite elastic modulus is more appropriate,

$$E = E_S/V_S + E_S V_L \tag{9.4}$$

where it is assumed that the two phases are under equivalent stress. For liquid phase sintered tungsten heavy alloys, Krock (14) found the upper bound was slightly lower than the measured values as shown in Figure 9.2. Alternatively, for tungsten-copper compacts (17) the measured values were in good agreement with the lower bound estimated from the uniform stress model of Equation (9.4).

Nakamura and Gurland (18) suggest a significantly more complex model for the elastic modulus;

$$E = [b_1^2 + b_1 b_2]/[b_1 + b_2 (1 - b_3)] \tag{9.5}$$

with

$$b_1 = E_L (1 - V_C) \tag{9.6}$$

$$b_2 = (E_S - E_L) (V_S - V_C)^{2/3} \tag{9.7}$$

$$b_3 = (V_S - V_C)^{1/3} \tag{9.8}$$

$$V_C = V_S C_{ss}. \tag{9.9}$$

The model has been applied to cemented carbides with good success (19). However, there is little justification offered for the complexity of this model.

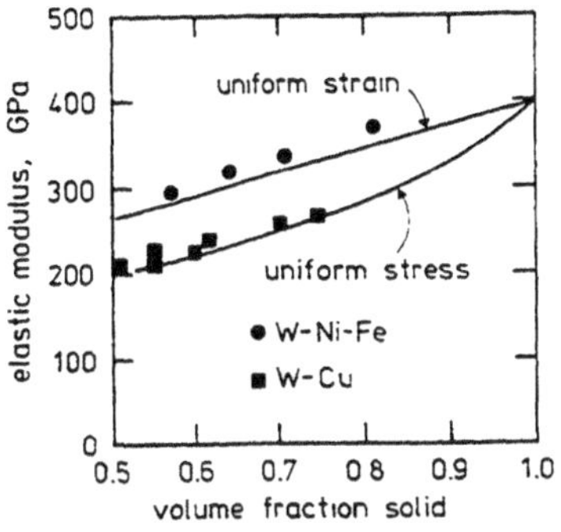

Figure 9.2 The elastic modulus versus volume fraction of solid for two tungsten-based materials (14,17) and the models given by Equations (9.3) and (9.4).

Spandoudakis and Young (20) have reported success in modelling the elastic modulus of particulate composites with a low contiguity using the following formula:

$$E = E_L + E_S V_S / [m/(m - 1) - V_S^{1/3}] \tag{9.10}$$

where the parameter m is the modulus ratio

$$m = E_S / E_L . \tag{9.11}$$

Their measurements agree with earlier findings for solid contents up to 60%. For glass spheres in an epoxy matrix, they report a slight decrease in elastic modulus with decreasing grain size.

3. *Strength*

Strength depends on microstructure and other factors such as the grain-matrix cohesion; large grain sizes and weak interfaces lower the strength (21-25). Thus, the strength of a two phase composite proves difficult to analyze mathematically based on just the microstructure. Strength typically is not systematic with any of the common microstructural parameters. At low matrix concentrations, the strength increases with the amount of matrix. For brittle grains in a soft matrix, there is an optimization point as illustrated in Figure 9.3 (8-10,15,26-28). Alternatively, for ductile grains in a harder matrix, the strength depends directly on the amount of matrix. Figure 9.4 demonstrates this latter behavior for a Fe-B alloy sintered at 1200°C. The strength increases with the amount of boron, although the ductility decreases simultaneously.

Various microstructure-property equations have been proposed to predict the behavior pattern illustrated in Figures 9.3 (28-30). In the

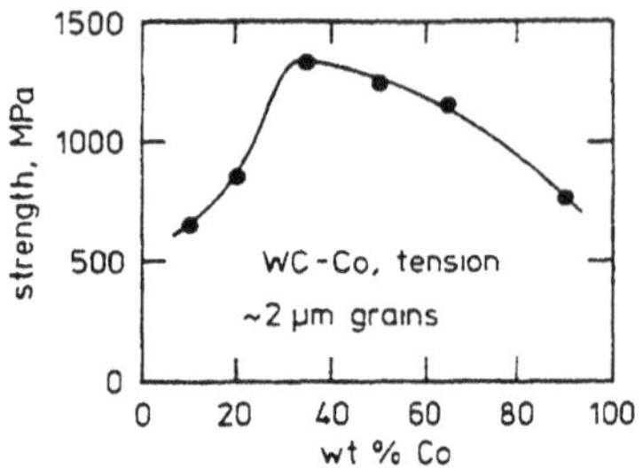

Figure 9.3 The strength of WC-Co in tension with a 2 µm grain size and various cobalt contents (15).

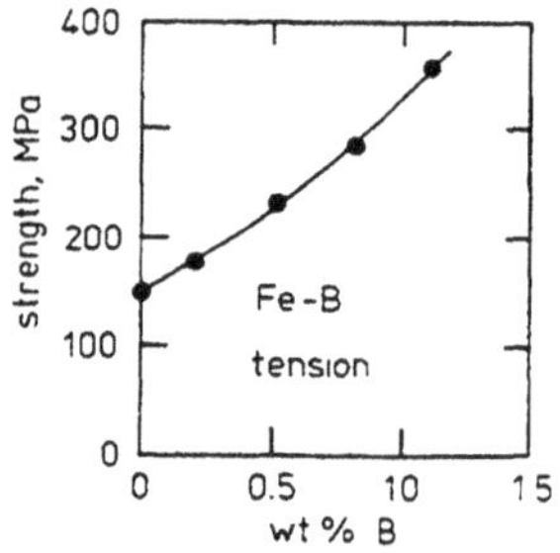

Figure 9.4 The tensile strength of Fe-B alloys versus the amount of boron, showing increasing strength with the amount of liquid.

cemented carbides, fracture is controlled by the largest grains; however, because of the microstructural scaling among the various features during liquid phase sintering, mean properties are a suitable gauge of the expected behavior (29). Zhenyao (30) derived an equation linking the strength σ to the grain size G and volume fractions as follows;

$$\sigma = k_o + (k_1/G)\,(V_S/V_L)^{1/2} - k_2\,(V_S V_L)^{1/2} \qquad (9.12)$$

where the k's are constants. The model is for an isolated microstructure, but seems to work for cemented carbides with high volume fractions of matrix in spite of this assumption. As an alternative, it has been proposed that Equation (9.1) would be applicable as a strength model, with strength substituted for hardness (13).

The low strength often observed at low matrix contents is thought to represent the strong sensitivity to defects rather than an actual microstructural dependence. In materials fabricated with low volume fractions of liquid

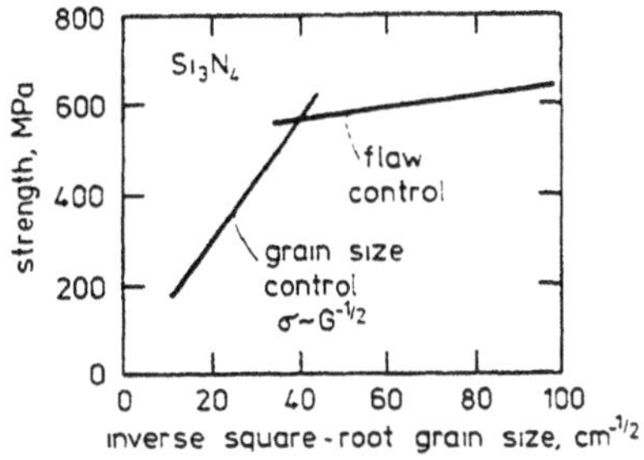

Figure 9.5 The change in strength behavior from flaw control to grain size control versus the grain size for silicon nitride materials (31).

phase, the defects (pores, inclusions, large grains) cause premature failure (10,31-35). The small amount of matrix fails to inhibit crack propagation initiated by these processing defects. At high volume fractions of matrix, there is more uniformity to the sintered structure and the matrix is more effective in suppressing crack growth (36). Figure 9.5 demonstrates this effect by showing the strength versus inverse square-root of the grain size for various silicon nitride materials (31). At small grain sizes the flaws control the strength, while at large grain sizes there is a true grain size effect. In the high volume fraction of matrix range, strength is determined by dispersion strengthening of the matrix. As the grain separation increases the grains provide less restriction to plastic flow and there is a declining strength. Thus, final stage grain growth must be controlled to optimize properties. This concept is supported by strength measurements conducted under compression, where the defect role is suppressed at high volume fractions of solid (10). In practice, covalent ceramics and cemented carbides are often hot isostatically pressed after sintering to heal the residual defects (37,38). The hot pressing treatment has a benefit only for the defect sensitive alloys with low volume fractions of liquid (34).

A high contiguity decreases the strength (10,25,39), as noted in Figure 9.6 for W-Ni-Fe alloys of increasing tungsten content (40). These data are for alloys in the sintered and heat treated condition, ranging from 80 to 99.5 wt.% tungsten. As the contiguity increases, the crack blunting ability of the matrix phase is decreased; thus, lower strengths and ductilities are observed. Finally, the effect of grain size is less certain. Various reports have shown an increase in grain size to raise the strength, decrease the strength, or exhibit a strength maximum at an intermediate grain size (8,10,12). These three behavior patterns are demonstrated in Figure 9.7 for WC-Co alloys with 6, 16, and 25% cobalt. At the lowest matrix content the strength increases with an increase in the grain size. However, at the high matrix contents the behavior is just the opposite. The intermediate cobalt content exhibits both characteristics. Subsequent analysis demonstrates that these patterns represent the underlying grain separation effect, where the mean grain separation is the dominant factor.

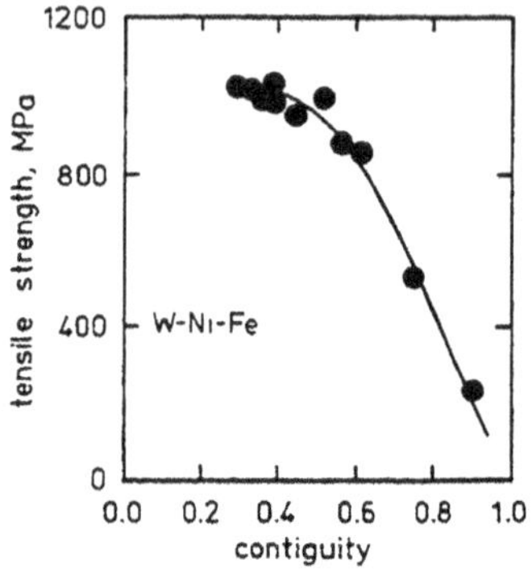

Figure 9.6 Contiguity effect on the tensile strength for various W-Ni-Fe alloys (40).

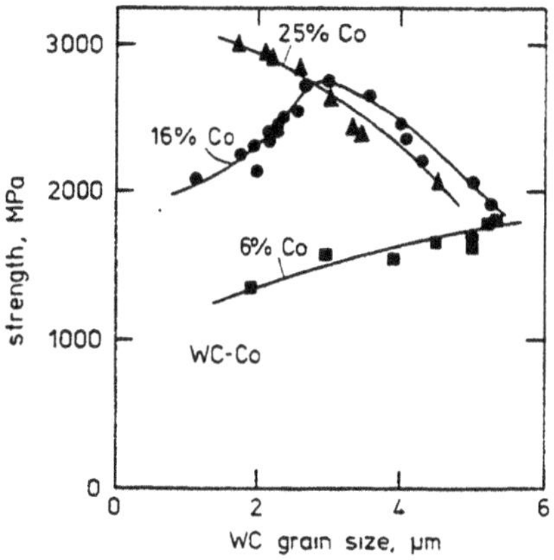

Figure 9.7 The strength of WC-Co alloys showing the combined influences of carbide grain size and cobalt content (8).

4. Ductility

Ductility has been measured on several liquid phase sintered materials. The effect of an increase in matrix content is to increase the grain separation. The corresponding change in the ductility depends on the matrix material. For a hard grain in a soft matrix, the ductility increases with the amount of matrix. At the very high tungsten contents in the heavy alloys, the ductility falls rapidly because of a rapid increase in the contiguity. In such cases the contiguity is the main determinant of ductility as noted

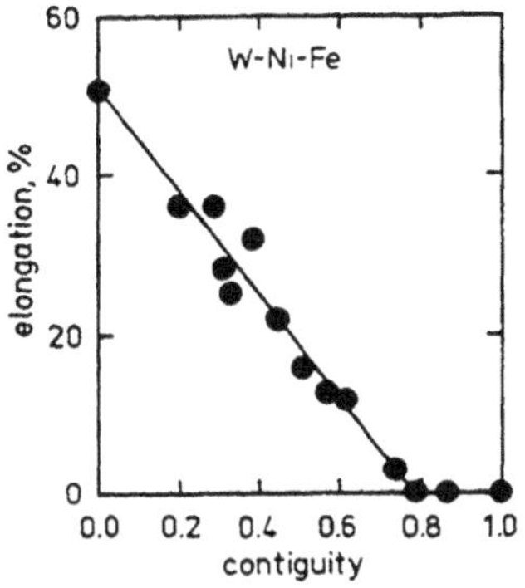

Figure 9.8 The elongation to failure of W-Ni-Fe alloys shown as a function of the solid phase contiguity (40).

earlier. For such cases the ductility at failure ε_f is limited by the contiguity C_{ss} as follows (25):

$$\varepsilon_f = K\,(1 - C_{ss}) \tag{9.13}$$

where K is a material constant. Ductility data for various W-Ni-Fe alloys are shown in Figure 9.8 (40). The ductility has an inverse dependence on the contiguity and becomes zero at a contiguity of approximately 0.8. Alternatively, for systems where the grains are more ductile than the matrix, ductility decreases as the amount of matrix increases. Figure 9.9 demonstrates this for Fe-Sn alloys, showing both strength and ductility versus the tin content (22). Here tin forms an embrittling grain boundary intermetallic phase between the ductile iron grains. The greater the quantity of tin, the lower the strength and ductility. No systematic investigation has been completed on the grain size and mean grain separation effects on ductility.

5. *Impact Toughness*

The impact toughness depends on the combination of strength and ductility. Accordingly, impact toughness shows a mixed dependence on the volume fraction of matrix (41,42). There is an optimal amount of matrix, similar to the effect noted for strength in Figures 9.3 and 9.7. The impact toughness decreases as the contiguity of the hard phase increases. Furthermore, toughness is sensitive to the segregation of embrittling impurities; thus, post-sintering heat treatments can have a significant effect.

C. Fracture

The fracture path in a liquid phase sintered material gives evidence of the microstructure effects on properties. During fracture the crack can pass

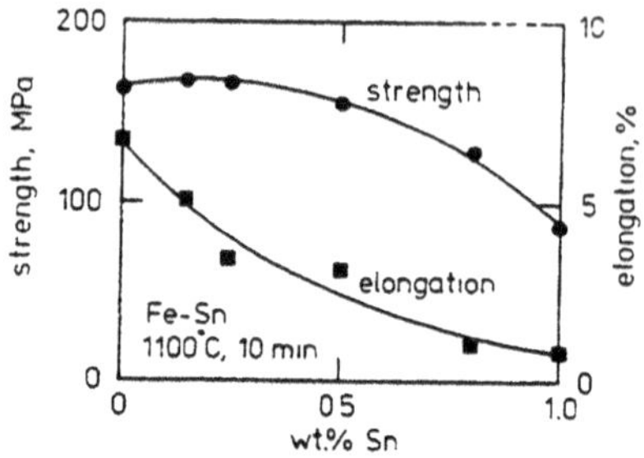

Figure 9.9 The strength and elongation of Fe-Sn compacts after sintering at 1100°C for 10 minutes, showing an embrittling effect of the tin (22).

along any of four possible paths as illustrated in Figure 9.10 (1,8). Fracture of the solid grains is often difficult because of their high hardness. The matrix provides toughness to the composite and often is a strong link. Alternatively, the solid grain boundaries are weak and the interface between the grain and the matrix can be of variable character (2,23). For high mechanical properties and resistance to fracture, the matrix-grain interface must be strong. This necessitates that the matrix-grain interface be free from precipitates, impurities, pores, and reaction products (21,38,43-47). The local stress transfer from the matrix to the hard grains is important for high mechanical properties. Under appropriate conditions, deformation of the grain is possible due to hydrostatic loading from the matrix (48). Such behavior is typically associated with excellent mechanical properties. The three scanning electron micrographs presented in Figure 9.11 contrast the fracture surfaces of low, moderate, and high ductility alloys. The brittle material (Figure 9.11a) shows separation of the matrix from the grain and exhibited low ductility, strength, and toughness (2). By contrast, the high ductility sample (Figure 9.11c) was heat treated after liquid phase sintering to remove interfacial segregation. The fracture surface evidences a greater quantity of cleavage and matrix failure as the mechanical properties improved.

The weak links associated with liquid phase sintered materials are typically the pores and points of solid-solid contact which give contiguity to the sintered product (9,25,39). The contiguity is necessary to prevent compact slumping during sintering. Unfortunately, these weak links dominate fracture and as a consequence it is desirable to have a low contiguity, small grain size, and large grain separation to increase the fracture toughness (32,49,50). A high volume fraction of matrix reduces the contiguity and increases the fracture toughness. If the interface between the grain and the matrix is clean, considerable deformation is possible prior to failure. Thus, it is favorable to maintain a low impurity level and to perform specific heat treatments to avoid impurity segregation (28,42,51,52). Often the liquid forming addition preferentially segregates to the grain boundaries and causes embrittlement (1,53,54). Likewise the formation of intermetallic phases which precipitate during cooling embrittles a liquid phase sintered material (44,55,56). Examination of material during failure shows that the largest grains are responsible for fracture initiation (28,29). Accordingly, a uniform

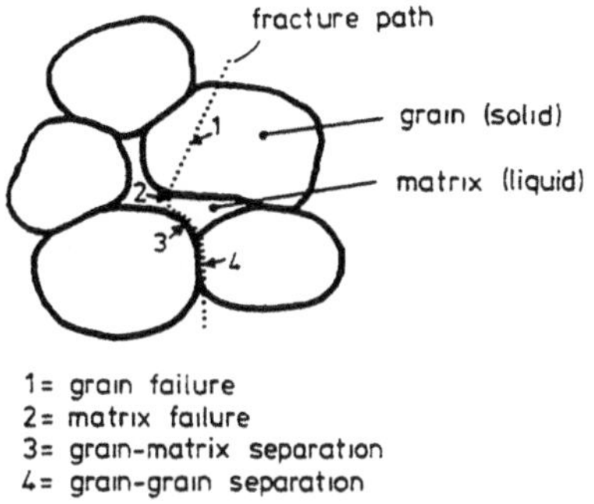

Figure 9.10 The four main fracture paths possible through a liquid phase sintered material.

grain size is desirable to avoid premature cracking. For these reasons the fracture behavior is very sensitive to the microstructure.

Several models have been proposed for linking fracture toughness with the microstructure. Almond (32) reviews several studies and notes the dominant role of the grain separation in improving fracture toughness. However, the models are largely empirical and can not be used to quantitatively predict how fracture toughness can be improved independent of changes in other properties dependent on the microstructure. A low contiguity, large matrix content, and thick matrix are useful characteristics. The fracture toughness (as measured by the stress intensity for crack propagation) has been correlated with the amount of cobalt, contiguity, and mean grain spacing for WC-Co alloys (8,18,28,39). In addition, the fracture toughness has been correlated to the inverse of the hardness for the cemented carbides (9,57,58). Figure 9.12 provides a demonstration of such a correlation for WC-Co. A variety of cobalt contents and grain sizes were used in constructing this plot of hardness and fracture energy. The inverse correlation between hardness and fracture is evident.

D. High Temperature Properties

The properties of liquid phase sintered materials at high temperatures are controlled by the additives used to form the liquid during sintering. As the test temperature is raised, properties like the strength, hardness, and elastic modulus generally decrease (59). At high temperatures the liquid phase can reform and cause catastrophic failure. Below the liquid formation temperature, accelerated creep results from the low melting temperature matrix (51,60). The resistance to creep failure at high temperatures increases as the amount of matrix phase and the test temperature are reduced (61).

Creep occurs preferentially in the matrix phase. Thus, creep resistance improves with smaller quantities of matrix, and higher contiguities. Figure 9.13 contrasts the temperature effects on fracture of silicon nitride samples fabricated with and without liquid phases (52). The strength of the liquid

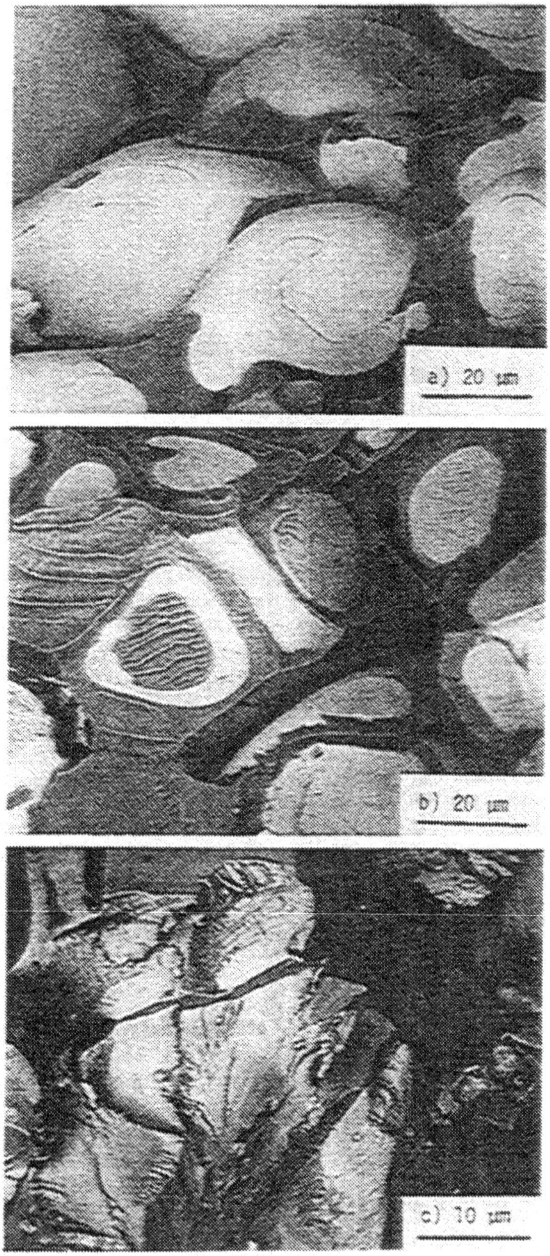

Figure 9.11 Fracture surfaces of W-Ni-Fe alloys illustrating the features associated with brittle failure (a), moderate ductility (b), and high ductility (c).

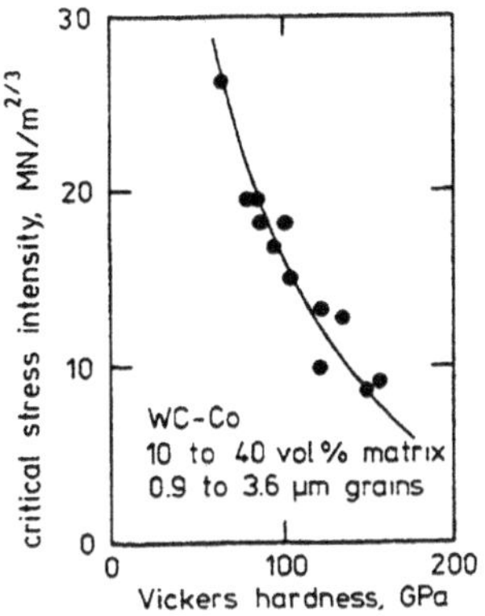

Figure 9.12 The critical stress intensity for crack propagation in WC-Co alloys versus the hardness (18).

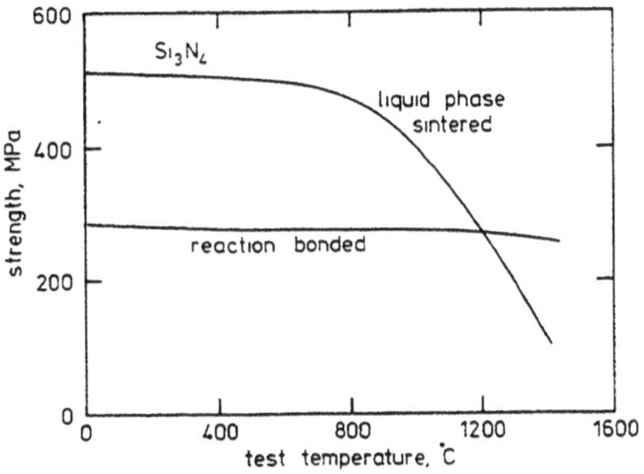

Figure 9.13 A contrast between the temperature dependent strengths of liquid phase sintered and reaction bonded (solid-state sintered) silicon nitrides (52).

phase sintered material is higher at low temperatures, but as the temperature rises, the strength falls comparatively rapidly. In such materials, failure is by void initiation in the matrix phase which has a high diffusion rate. The voids nucleate and grow under an applied stress. Failure occurs when the voids link together. Void formation is easier as the quantity of matrix increases. For this reason, transient liquid phase sintering is more attractive for materials used at high temperatures (31). Stress concentrations in the microstructure are quite detrimental; thus, inclusions, large grains, and thick matrix layers are to be avoided.

E. Thermal Properties

The thermal behavior of liquid phase sintered materials is similar to that of other multiple phase composites. As a first approximation to the thermal properties, a rule of mixtures is applicable. For example, the thermal expansion coefficient α will depend on the thermal expansion coefficients of the two phases as follows:

$$\alpha = \alpha_L V_L + \alpha_S V_S. \tag{9.14}$$

However, a difference in thermal expansion coefficients between the two phases results in an interfacial stress. Various models have been proposed to account for the interfacial stress. The model by Fahmy and Ragai (62) assumes spherical grains dispersed in a matrix, giving a complex dependence on the amount of each phase, their elastic moduli, and Poisson ratios;

$$\alpha = \alpha_L - 3\,(\alpha_L - \alpha_S)(1 - \nu_L)\,V_S / [2\,(E_L/E_S)(1 - 2\nu_S\,V_L + 2\,V_S\,(1 - 2\,\nu_L) + (1 + \nu_L)] \tag{9.15}$$

where the subscripts S and L indicate the solid and matrix (liquid) phases, respectively. In Equation (9.15), α represents the thermal expansion coefficient, ν represents Poisson's ratio, E is the elastic modulus, and V represents the volume fraction. Other theories, assume more complex strain states and grain shapes (63-65). Unfortunately, the complexity of the mathematical formulations is not appealing since several of the properties must be estimated and the assumed microstructure is fairly simple.

Thermal conductivity parallels the electrical conductivity, as discussed in the next section.

F. Electrical Properties

The electrical conductivity of a two phase composite has a complex dependence on the characteristics of the two phases, grain shape, and connectivity. Nazare et al. (66) have summarized the equations treating electrical conductivity variations with structure and quantity of each phase. In typical cases involving electrical applications, the grains are less conductive than the matrix (for example W-Cu). Thus, as shown in Figure 9.14, the conductivity decreases as the amount of solid increases (67). This figure shows the electrical conductivity of W-Cu (as a percent of annealed copper) versus the tungsten content. For high electrical conductivity it has been found that coarse grains, high sintered density, and low solid contents are the dominant factors. Also, solubilities affect conductivity. Both the electrical and thermal conductivities increase for W-Cu materials as the hardness increases (68).

A typical application for a liquid phase sintered material would be as an electrical contact. A common problem is with burn-off of the contact during arcing. The rate of material loss increases as the grain size increases (68). The coarse grain materials have a lower hardness and suffer from higher vapor losses during arcing. One mode of failure for contact materials is by

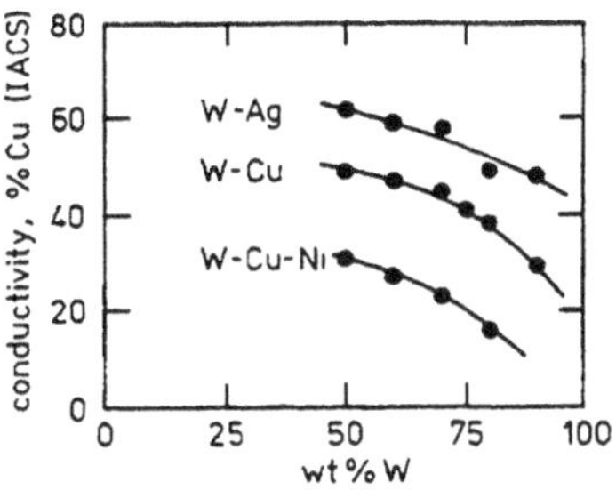

Figure 9.14 The relative electrical conductivity of tungsten-based alloys versus the tungsten content (67).

preferential evaporation of the matrix phase. A small grain separation (small grain size) provides a higher capillary pressure on the matrix at the arcing surface and less burn-off. However, the toughness of the composite is lowest with the small grain size, consequently cracking due to thermal stresses is observed. The failure of W-Ag electrical contacts under arcing conditions has been related to the microstructure (69). A high fracture toughness is desirable to inhibit crack propagation under arcing conditions. For W-Ag materials the fracture toughness K_{IC} has been related to the microstructure as follows:

$$K_{IC}^2 = V_L \lambda / C_{ss} \tag{9.16}$$

where λ is the mean separation between tungsten grains. Cracking is most probable with a low fracture toughness; thus, a high hardness, large grain separation, and low contiguity give the best resistance to cracking during arcing.

G. Wear Behavior

Wear behavior of liquid phase sintered materials is a major concern because of their widespread applications in drilling, earth moving, tooling, machining, hardfacing, and metalforming operations. Carbides, nitrides, borides, and some oxides are used in severe stress situations where wear is often the primary concern. Obviously the grains of these liquid phase sintered materials are quite hard. There are several possible wear mechanisms (diffusion, adhesion, attrition, abrasion, and fatigue). These wear mechanisms couple with a dependence on corrosion, stress, temperature, geometry, and opposing materials to determine the actual wear rate (70).

In the usual case, the grains of the liquid phase sintered material are harder than the material causing the wear. In this condition, microstructure plays a significant role in determining the wear rate (71,72). The usual behavior is for the amount of wear to increase with the applied load. Furthermore, abrasion wear increases as the mean separation between the grains increases and as the hardness decreases. This is primarily due to preferential removal of the matrix. Under impact conditions the wear rate is minimized by an intermediate hardness because of the concomitant decrease in fracture toughness at high hardnesses.

An investigation by Chermant and Osterstock (28,39) found that wear generally increased with the fracture toughness. Consequently, they suggested that composites with low volume fractions of matrix showed best wear resistance with fine grain sizes. Alternatively, at high volume fractions of matrix a coarse grain size was superior. Overall, best wear resistance was found with low volume fractions of matrix and small grain sizes (less than 0.7 μm). In situations involving corrosion, wear is also dependent on the matrix corrosion rate.

H. Magnetic Properties

Liquid phase sintered materials are used for both soft and hard magnet applications. For both applications high sintered densities are desired. Hard magnets have high residual magnetizations and high coercive forces. To obtain these characteristics, the sintered product requires a low defect population in a single phase, stoichiometric compound (73). A typical compound for hard magnetic applications is $SmCo_5$ sintered with a liquid phase from hyperstoichiometric powder.

The hyperstoichiometric samarium cobalt powder is mixed with excess samarium to form a transient liquid during sintering. In sintering samarium cobalt magnets, too low a sintering temperature fails to give densification. Alternatively, too high a sintering temperature gives samarium loss by evaporation. As a consequence a second phase will form in the microstructure. Thus, optimal magnetic behavior dictates control over the grain size, density, defect structure, and stoichiometry (73-76).

The soft magnetic materials are often used in alternating current fields where a high permeability, low coercive force, and low eddy current loss are desired. Materials like Fe-P and Fe-Si are liquid phase sintered for these applications (77-82). Again a high sintered density is desired, with spherical pores and a large grain size. Liquid phase sintering gives a higher final density, rapid grain growth, and smoother pore shape; thus, it is an attractive approach for sintering soft magnetic materials.

I. Summary

Mechanical properties are the best documented properties of liquid phase sintered materials. Fracture studies have lead to a concept of weak links with a hierarchy of factors affecting the observed behavior. For most cases, the solid is also the harder phase; thus, the ductile matrix controls the observed properties. To assure good properties, it is necessary to avoid precipitates, segregated films, pores, and microstructural inhomogeneities. The various properties have differing sensitivities to the sintered microstructure. A property like the elastic modulus is sensitive only to the amount of each phase and the connectivity. Hardness, is more sensitive to the microstructure, but there are fairly straightforward controlling relations. Alternatively, ductility is very sensitive to both the microstructure and subtle factors like impurity segregation. For the more sensitive properties like ductility, optimal microstructures are observed. The main factors are the amount of each phase, grain size, contiguity, and grain separation.

Not all liquid phase sintered materials exhibit acceptable properties, even when processed to give a high sintered density. Figure 9.9 gives an example of such behavior. This figure shows the strength versus amount of tin addition for sintered Fe-Sn alloys. A decrease in the strength occurs as the amount of tin increases because of grain boundary precipitation of an intermetallic compound.

J. References

1. R. M. German and J. E. Hanafee, "Processing Effects on Toughness for Liquid Phase Sintered W-Ni-Fe," *Processing of Metal and Ceramic Powders*, R. M. German and K. W. Lay (eds.), The Metallurgical Society, Warrendale, PA, 1982, pp.267-282.
2. R. M. German, J. E. Hanafee, and S. L. DiGiallonardo, "Toughness Variation with Test Temperature and Cooling Rate for Liquid Phase Sintered W-3.5Ni-1.5Fe," *Metall. Trans. A*, 1984, vol.15A, pp.121-128.
3. Z. Hashin, "Analysis of Composite Materials - A Survey," *J. Appl. Mech.*, 1983, vol.50, pp.481-505.
4. H. Doi, *Elastic and Plastic Properties of WC-Co Composite Alloys*, Freund Scientific Publications, Tel-Aviv, Israel, 1974.
5. H. Fischmeister and B. Karlsson, "Plastizitatseigenschaften Grob-Zweiphasiger Werkstoffe," *Z. Metallkde.*, 1977, vol.68, pp.311-327.
6. R. K. Viswanadham and J. D. Venables, "A Simple Method for Evaluating Cemented Carbides," *Metall. Trans. A*, 1977, vol.8A, pp.187-191.
7. G. Grathwohl and R. Warren, "The Effect of Cobalt Content on the Microstructure of Liquid-Phase Sintered TaC-Co Alloys," *Mater. Sci. Eng.*, 1974, vol.14, pp.55-65.
8. J. Gurland and P. Bardzil, "Relation of Strength, Composition, and Grain Size of Sintered WCCo Alloys," *Trans. TMS-AIME*, 1955, vol.203, pp.311-315.
9. R. Warren and B. Johannesson, "The Fracture Toughness of Hardmetals," *Inter. J. Refractory Hard Met.*, 1984, vol.3, pp.187-191.
10. H. E. Exner and J. Gurland, "A Review of Parameters Influencing Some Mechanical Properties of Tungsten Carbide-Cobalt Alloys," *Powder Met.*, 1970, vol.13, pp.13-31.
11. F. V. Lenel, *Powder Metallurgy Principles and Applications*, Metal Powder Industries Federation, Princeton, NJ, 1980, pp.383-400.
12. V. K. Sarin, "Cemented Carbide Cutting Tools," *Advances in Powder Technology*, G. Y. Chin (ed.), American Society for Metals, Metals Park, OH, 1982, pp.253-288.
13. J. Gurland, "A Structural Approach to the Yield Strength of Two-Phase Alloys with Coarse Microstructures," *Mater. Sci. Eng.*, 1979, vol.40, pp.59-71.
14. R. H. Krock, "Elastic and Plastic Deformation of Dispersed Phase Liquid Phase Sintered Tungsten Composite Materials," *Metals for the Space Age*, F. Benesovsky (ed.), Metallwerk Plansee, Reutte, Austria, 1965, pp.256-275.
15. C. Nishimatsu and J. Gurland, "Experimental Survey of the Deformation of the Hard-Ductile Two-Phase Alloy System WC-Co" *Trans. ASM*, 1960, vol.52, pp.469-484.
16. B. Paul, "Prediction of Elastic Constants of Multiphase Materials," *Trans TMS-AIME*, 1960, vol.218, pp.36-41.
17. H. Krock, "Some Comparisons Between Fiber-Reinforced and Continuous Skeleton Tungsten-Copper Composite Materials," *J. Mater.*, 1966, vol.1,

pp.278-292.
18. M. Nakamura and J. Gurland, "The Fracture Toughness of WCCo Two Phase Alloys - A Preliminary Model," *Metall. Trans. A,* 1980, vol.11A, pp.141-146.
19. J. R. Pickens and J. Gurland, "The Fracture Toughness of WCCo Alloys Measured on Single-Edge Notched Beam Specimens Précracked by Electron Discharge Machining," *Mater. Sci. Eng.,* 1978, vol.33, pp.135-142.
20. J. Spanoudakis and R. J. Young, "Crack Propagation in a Glass Particle-Filled Epoxy Resin. Part 1. Effect of Particle Volume Fraction and Size," *J. Mater. Sci.,* 1984, vol.19, pp.473-486.
21. D. V. Edmonds and P. N. Jones, "Interfacial Embrittlement in Liquid-Phase Sintered Tungsten Heavy Alloys," *Metall. Trans. A,* 1979, vol.10A, pp.289-295.
22. T. R. Moules and C. A. Calow, "Studies on the Activation-Sintering of Iron Powder," *Powder Met.,* 1972, vol.15, pp.55-66.
23. J. Spandoukakis and R. J. Young, "Crack Propagation in a Glass Particle-Filled Epoxy Resin. Part 2. Effect of Particle-Matrix Adhesion," *J. Mater. Sci.,* 1984, vol.19, pp.487-496.
24. C. Li and R. M. German, "The Properties of Tungsten Processed by Chemically Activated Sintering," *Metall. Trans. A,* 1983, vol.14A, pp.2031-2041.
25. K. S. Churn and R. M. German, "Fracture Behavior of W-Ni-Fe Heavy Alloys," *Metall. Trans. A,* 1984, vol.15A, pp.331-338.
26. L. Albano-Muller, F. Thummler, and G. Zapf, "High-Strength Sintered Iron-Base Alloys by Using Transition Metal Carbides," *Powder Met.,* 1973, vol.16, pp.236-256.
27. J. L. Chermant and F. Osterstock, "Elastic and Plastic Characteristics of WC-Co Composite Materials," *Powder Met. Inter.,* 1979, vol.11, pp.106-109.
28. J. L. Chermant, A. Deschanvres, and F. Osterstock, "Factors Influencing the Rupture Stress of Hardmetals," *Powder Met.,* 1977, vol.20, pp.63-69.
29. J. L. Chermant, M. Coster, G. Hautier, and P. Schaufelberger, "Statistical Analysis of the Behaviour of Cemented Carbides under High Pressure," *Powder Met.,* 1974, vol.17, pp.85-102.
30. T. Zhenyao, "A New Statistical Relation Between the Strength and the Microstructural Parameters," *Mater. Sci. Eng.,* 1982, vol.56, pp.73-85.
31. J. Lorenz, J. Weiss, and G. Petzow, "Dense Silicon Nitride Alloys: Phase Relations and Consolidation, Microstructure and Properties," *Advances in Powder Technology,* G. Y. Chin (ed.), American Society for Metals, Metals Park, OH, 1982, pp.289-308.
32. E. A. Almond, "Deformation Characteristics and Mechanical Properties of Hardmetals," *Science of Hard Materials,* R. K. Viswanadham, D. J. Rowcliffe and J. Gurland (eds.), Plenum Press, New York, NY, 1983, pp.517-557.
33. S. Amberg and H. Doxner, "Porosity in Cemented Carbides," *Powder Met.,* 1977, vol.20, pp.1-10.
34. C. Chatfield, "Comments on Microstructure and the Transverse Rupture Strength of Cemented Carbides," *Inter. J. Refractory Hard Met.,* 1985, vol.4, pp.48.
35. P. B. Anderson, "Hartmetalle erhohter Zahigkeit," *Planseeber. Pulvermet.,* 1967, vol.15, pp.180-186.
36. L. LeRoux, "Microstructure and Transverse Rupture Strength of Cemented Carbides," *Inter. J. Refract. Hard Met.,* 1984, vol.3,

pp.99-100.
37. U. Engel and H. Hubner, "Strength Improvement of Cemented Carbides by Hot Isostatic Pressing (HIP)," *J. Mater. Sci.*, 1978, vol.13, pp.2003-2012.
38. S. Amberg, E. A. Nylander, and B. Uhrenius, "The Influence of Hot Isostatic Pressing on the Porosity of Cemented Carbide," *Powder Met. Inter.*, 1974, vol.6, pp.178-180.
39. J. L. Chermant and F. Osterstock, "Fracture Toughness and Fracture of WCCo Composites," *J. Mater. Sci.*, 1976, vol.11, pp.1939-1951.
40. R. M. German and L. L. Bourguignon, "Analysis of High Tungsten Content Heavy Alloys," *Powder Metallurgy in Defense Technology*, vol.6, C. L. Freeby and W. J. Ullrich (eds.), Metal Powder Industries Federation, Princeton, NJ, 1985, pp.117-131.
41. F. Osterstock, "Impact Behaviour of Tungsten Carbide-Cobalt Alloys," *Inter. J. Refractory Hard Metals*, 1983, vol.2, pp.116-120.
42. F. E. Sczerzenie and H. C. Rogers, "Hydrogen Embrittlement of Tungsten Base Heavy Alloys," *Hydrogen in Metals*, I. M. Bernstein and A. W. Thompson (eds.), American Society for Metals, Metals Park, OH, 1974, pp.645-655.
43. S. B. Luyckx, "Role of Inclusions in the Fracture Initiation Process in WC-Co Alloys," *Acta Met.*, 1975, vol.23, pp.109-115.
44. C. J. Li and R. M. German, "Enhanced Sintering of Tungsten - Phase Equilibria Effects on Properties," *Inter. J. Powder Met. Powder Tech.*, 1984, vol.20, pp.149-162.
45. B. C. Muddle, "Interphase Boundary Precipitation in Liquid Phase Sintered W-Ni-Fe and W-Ni-Cu Alloys," *Metall. Trans. A*, 1984, vol.15A, pp.1089-1098.
46. B. C. Muddle and D. V. Edmonds, "Interfacial Segregation and Embrittlement in Liquid Phase Sintered Tungsten Alloys," *Metal Sci.*, 1983, vol.17, pp.209-218.
47. K. S. Churn and D. N. Yoon, "Pore Formation and its Effect on Mechanical Properties in W-Ni-Fe Heavy Alloy," *Powder Met.*, 1979, vol.22, pp.175-178.
48. L. Ekbom, "Microstructural Study of the Deformation and Fracture Behavior of a Sintered Tungsten-Base Composite," *Modern Developments in Powder Metallurgy*, vol.14, H. H. Hausner, H. W. Antes and G. D. Smith (eds.), Metal Powder Industries Federation, Princeton, NJ, 1981, pp.177-188.
49. C. T. Peters, "The Relationship Between Palmqvist Indentation Toughness and Bulk Fracture Toughness for Some WC-Co Cemented Carbides," *J. Mater. Sci.*, 1979, vol.14, pp.1619-1623.
50. K. S. Cherniavsky, "Stereology of Cemented Carbides," *Sci. Sintering*, 1982, vol.14, pp.1-12.
51. J. E. Marion, A. G. Evans, M. D. Drory, and D. R. Clarke, "High Temperature Failure Initiation in Liquid Phase Sintered Materials," *Acta Met.*, 1983, vol.31, pp.1445-1457.
52. R. L. Tsai and R. Raj, "Creep Fracture in Ceramics Containing Small Amounts of a Liquid Phase," *Acta Met.*, 1982, vol.30, pp.1043-1058.
53. P. Lindskog, "The Effect of Phosphorus Additions on the Tensile, Fatigue, and Impact Strength of Sintered Steels Based on Sponge Iron," *Powder Met.*, 1973, vol.16, pp.374-386.
54. R. L. Hodson and N. M. Parikh, "Cemented Carbides with High-Melting Binders II: Ternary Equilibrium Systems," *Inter. J. Powder Met.*, 1967, vol.3, no.3, pp.31-40.

55. K. S. Hwang and R. M. German, "High Density Ferrous Components by Activated Sintering," *Processing of Metal and Ceramic Powders*, R. M. German and K. W. Lay (eds.), The Metallurgical Society, Warrendale, PA, 1982, pp.295-310.
56. R. M. German and B. H. Rabin, "Enhanced Sintering Through Second Phase Additions," *Powder Met.*, 1985, vol.28, pp.7-12.
57. T. Sadahiro and S. Takatsu, "A New Precracking Method for Fracture Toughness Testing of Cemented Carbides," *Modern Developments in Powder Metallurgy*, vol.14, H. H. Hausner, H. W. Antes and G. D. Smith (eds.), Metal Powder Industries Federation, Princeton, NJ, 1981, pp.561-572.
58. S. Singh, "Palmqvist Toughness of Cemented Carbide Alloys," *Inter. J. Refractory Hard Met.*, 1985, vol.4, pp.27-30.
59. P. Schwarzkopf and R. Kieffer, *Cemented Carbides*, MacMillan Co., New York, NY, 1960.
60. M. D. Thouless and A. G. Evans, "Nucleation of Cavities During Creep of Liquid Phase Sintered Materials," *J. Amer. Ceramic Soc.*, 1984, vol.67, pp.721-727.
61. J. T. Smith and J. D. Wood, "Elevated Temperature Compressive Creep Behavior of Tungsten Carbide-Cobalt Alloys," *Acta Met.*, 1968, vol.16, pp.1219-1226.
62. A. A. Fahmy and A. N. Ragai, "Thermal Expansion Behavior of Two-Phase Solids," *J. Appl. Phys.*, 1970, vol.41, pp.5108-5111.
63. K. Wakashima, M. Otsuka, and S. Umekawa, "Thermal Expansions of Heterogeneous Solids Containing Aligned Ellipsoidal Inclusions," *J. Composite Mater.*, 1974, vol.8, pp.391-404.
64. L. I. Tuchinskii, "Thermal Expansion of Composites with a Skeletal Structure," *Soviet Powder Met. Metal Ceram.*, 1983, vol.22, pp.659-664.
65. S. J. Feltham, B. Yates, and R. J. Martin, "The Thermal Expansion of Particulate-Reinforced Composites," *J. Mater. Sci.*, 1982, vol.17, pp.2309-2323.
66. S. Nazare, G. Ondracek, and F. Thummler, "Relations Between Stereometric Microstructure and Properties of Cermets and Porous Materials," *Modern Developments in Powder Metallurgy*, vol.5, H. H. Hausner (ed.), Plenum Press, New York, NY, 1971, pp.171-186.
67. N. C. Kothari, "Factors Affecting Tungsten-Copper and Tungsten-Silver Electrical Contact Materials," *Powder Met. Inter.*, 1982, vol.14, pp.139-159.
68. G. H. Gessinger and K. N. Melton, "Burn-off Behaviour of W-Cu Contact Materials in an Electric Arc," *Powder Met. Inter.*, 1977, vol.9, pp.67-72.
69. G. J. Witter and W. R. Warke, "A Correlation of Material Toughness, Thermal Shock Resistance, and Microstructure of High Tungsten, Silver-Tungsten Composite Materials," *IEEE Trans. Parts Hybrids Packaging*, 1975, vol.11, pp.21-29.
70. A. Ball and A. W. Paterson, "Microstructural Design of Erosion Resistant Hard Materials," *Proceedings Eleventh International Plansee Seminar*, vol.2, H. Bildstein and H. M. Ortner (eds.), Metallwerk Plansee, Reutte, Austria, 1985, pp.377-391.
71. J. Larsen-Basse, "Effect of Composition, Microstructure, and Service Conditions on the Wear of Cemented Carbides," *J. Metals*, 1983, vol.35, no.11, pp.35-42.
72. J. Larsen-Basse, "Wear of Hard-Metals in Rock Drilling: A Survey of the Literature," *Powder Met.*, 1973, vol.16, pp.1-32.

73. Y. G. Bogatin, "Effect of Phase and Structural Transformations Occurring During Liquid-Phase Sintering on the Magnetic Properties of Samarium Cobalt Magnets," *Soviet Powder Met. Metal Ceram.*, 1978, vol.17, pp.393-398.
74. G. H. Gessinger and E. De Lamotte, "Der Sintermechanismus von Samarium-Kobalt-Legierungen," *Z. Metallkde.*, 1973, vol.64, pp.771-775.
75. P. J. Jorgensen and R. W. Bartlett, "Liquid-Phase Sintering of $SmCo_5$," *J. Appl. Phys.*, 1973, vol.44, pp.2876-2880.
76. P. J. Jorgensen and R. W. Bartlett, "Solid-Phase Sintering of $SmCo_5$," *J. Less-Common Metals*, 1974, vol.37, pp.205-212.
77. G. Jangg, M. Drozda, H. Danninger, H. Wibbeler, and W. Schatt, "Sintering Behavior, Mechanical and Magnetic Properties of Sintered Fe-Si Materials," *Inter. J. Powder Met. Powder Tech.*, 1984, vol.20, pp.287-300.
78. G. Jangg, M. Drozda, H. Danninger, and G. Eder, "Magnetic Properties of Sintered Fe-P Materials," *Powder Met. Inter.*, 1984, vol.16, pp.264-267.
79. F. Frehn and W. Hotop, "Effect of Small Boron Contents on the Properties of Compacts Prepared by Vacuum Sintering," *Symposium on Powder Metallurgy*, Special Report 58, Iron and Steel Institute, London, UK, 1956, pp.137-143.
80. K. H. Moyer, "The Effects of Phosphorus on the Properties of Iron Alloys for Magnetic Applications," *Prog. Powder Met.*, 1981, vol.37, pp.81-98.
81. W. Rutkowski and B. Weglinski, "The Influence of Silicon Additions on the Magnetic Properties of Iron Sinters," *Planseeber. Pulvermetall.*, 1979, vol.27, pp.162-177.
82. W. Rutkowski and B. Weglinski, "Phosphorus and its Influence on Properties of Magnetically Soft Sinters," *Planseeber. Pulvermetall.*, 1980, vol.28, pp.39-55.

CHAPTER TEN

Applications for Liquid Phase Sintering

A. Introduction

There are several materials commercially processed by liquid phase sintering. Additionally, there are several different industrial processing techniques and property combinations based upon liquid phase sintering. Because of this diversity of materials, processing cycles, and properties, this final chapter on applications will be overview in nature. The approach is to give a brief outline of the uses for liquid phase sintered materials. In this manner the several topics introduced in the earlier chapters are placed in perspective by the diversity of applications.

Table 10.1 lists several applications for liquid phase sintered materials with some example compositions and references for further information (1-40). It is evident from this table that liquid phase sintering has a broad spectrum of uses. In this chapter, four material types have been singled out for extended discussion; ferrous alloys, cemented carbides, tungsten heavy alloys, and silicon nitride compounds. These represent metallic, metal-ceramic, and ceramic materials. Additionally, these materials are used because of their unique properties which include magnetic, mechanical, friction, cutting, grinding, machining, radiation, and high temperature characteristics. As is evident there are several other materials processed by liquid phase sintering techniques which will not be detailed here. It is anticipated that the indicated references in Table 10.1 will aid the interested reader in a study of individual applications.

B. Ferrous Systems

Great interest exists in the liquid phase sintering of iron-based powder compacts (36,41,42). In examining this field, a distinction must be made between the two main processing goals of densification and dimensional control. Densification is desired to eliminate pores and obtain the highest properties, such as strength, ductility, toughness, magnetic permeability, and fatigue resistance. Here dimensional change is sought during sintering in order to remove pores. Alternatively, considerable liquid phase sintering of iron-based compositions is performed with the desire for no dimensional change during sintering. This is predominantly for applications involving complex shapes such as for mechanical components.

TABLE 10.1

Example Applications for Liquid Phase Sintering

Application	Example Compositions	References
Aerospace	Be-Si, Ti alloys, Ni-base alloys	1-3
Bearings	Cu-Sn, Al-Pb, Cu-Sn-Sb	4-6
Cutting Tools	WC-Co, TiC-Mo-Ni, tool steels	7-10
Dental	Ag-Cu-Sn-Hg	11
Electrical Capicators	$BaTiO_3$-LiF-MgO, $SrTiO_3$-SiO_2	12,13
Electrical Contacts	CdO-Ag, W-Ag-Ni, W-Cu-P	4,14,15
Filters	Cu-Sn, stainless steel-B	4,16
Friction Materials	Fe-Al_2O_3-C, Fe-C-Cu-Sn, Cu-Sn-SiO_2	4,17
Grinding Materials	diamond-metal, WC-Co, Al_2O_3-glass	4,18
High Temperature	Si_3N_4-MgO, SiC-B	9,19
Metalworking	Si_3N_4-Y_2O_3, tool steel, WC-TiC-Co	9,20-22
Nuclear	UO_2Al, UO_2-Al_2O_3, W-Ni-Fe	4,23,24
Permanent Magnets	$SmCo_5$-Sm, Fe-Al-Ni-Co-Cu	4,25-27
Porcelain	K_2O-Al_2O_3-SiO_2	28
Refractories	MgO-CaO-SiO_2, W-Cr-Al_2O_3 Al_2O_3-MgO-SiO_2	29-32
Soft Magnets	Fe-P, Fe-Si	33-34
Structural Components	Fe-Cu-C, Fe-Cr_3C_2, Fe-Mn-Cu-Si-Cu	35-37
Wear Materials	Co-base alloys, TiC-Fe, WC-Co	38-40

One important class of liquid phase formers are the nonmetals which have eutectics with iron, such as carbon, boron, and phosphorous. The addition of these to iron improves the strength of the sintered product. Additionally, copper is used because it melts during the traditional sintering cycle. Often copper is combined with carbon, phosphorus, or boron to control dimensions during sintering while attaining a high sintered strength. Other additions explored in the past include tin, titanium, sulfur, carbides, silicon, and manganese (34,37,43-50). In general, multiple component systems are used to obtain the several desirable features simultaneously. Most of the interest has been in the structural properties, although the magnetic characteristics have also been of great concern.

An increase in the amount of liquid phase typically results in strengthening, often causing a decreased ductility (51). However, there is

TABLE 10.2

Examples of Liquid Phase Sintered Ferrous Alloys

Alloy System wt.%	Sintering Cycle	Strength MPa	Elongation %
Fe-3Cu-2Sn	1000°C,1h	300	3
Fe-4Mo-1.1B	1200°C,1h	500	0
Fe-8Ni-1.1B	1100°C,1h	390	2
Fe-8Cu	1100°C,3h	430	20
Fe-3.5Ti	1330°C,2h	440	24
Fe-0.7C-0.5P	1120°C,1h	480	3
Fe-2Cu-0.8C	1175°C,0.5h	500	1
Fe-3Mo-4Ni-0.1P-0.1C-0.2B	1185°C,1h	720	4
Fe-5Cr_3C_2	1290°C,1.5h	890	2

greater dimensional change as the amount of liquid increases. Densities of 96 to 99% of theoretical have been attained with relatively small concentrations of additive. This is most attractive since coarse iron particle sizes have proven successful. In most all applications, temperature proves to be a major process control parameter. In the as-sintered condition, coarse iron powders with appropriate chemical additions have given strengths up to 800 MPa. Unfortunately, the ductilities are low, often below 2% elongation at these strength levels. Furthermore, several of the liquid forming additives do not aid strength in spite of their densification benefits. Tin is such an example. It leads to a low grain boundary strength and easy fracture along the tin-rich boundaries.

Table 10.2 gives the structural properties of several iron-based alloys processed by liquid phase sintering (36). Several options are available for attaining sintered strengths of at least 500 MPa. The attainment of higher strengths requires considerable attention to the processing cycle, additive homogeneity, and post-sintering heat treatment. In general, liquid phases are most beneficial if they form by a eutectic reaction. Additives based on nonmetallic phases perform well, including B, P, C, and Si. Also, ferrite stabilizers often prove to be attractive additives. The optimal composition depends on the sintering conditions, powder characteristics, microstructure, and alloying.

Iron-base liquid phase sintered alloys are used in components for automobiles, household appliances, farming equipment, office machines, and electrical motors. The general concern is with the mechanical strength and dimensional control. Besides mechanical components, liquid phase sintering is applied to the fabrication of iron-based friction materials (such as disk brakes and clutches), bearings, forging feedstock, and magnetic components (4).

C. Cemented Carbides

The cemented carbides are also known as hardmetals. They are composed of transition metal compounds of carbon in matrix phases based typically on cobalt, iron, or nickel. The classic example of a cemented carbide is WC-Co (52-55). Similar compounds involving borides or nitrides are under active development to extend the capabilities of hardmetals.

In the cemented carbides the properties are largely dictated by the composition, although there is considerable microstructural effect (7,56,57). Table 10.3 provides a general composition guide to the cemented carbides. The main factors affecting performance are the carbide composition, grain size, volume fraction of matrix, porosity, matrix phase composition, and internal strains. The general interest has been in obtaining high sintered hardnesses, with a more recent emphasis on toughness and fracture resistance.

Liquid phase sintering is the only proven technique for producing the cemented carbides (58-60). Densification is rapid because of a favorable solubility ratio and good wetting. The sintering cycle, especially the sintering temperature, effects the microstructure and is a main process control parameter (61-63). The sintering process is preceded by milling of the powders to reduce the mean particle size and obtain a homogeneous mixture of ingredients. The milled powders are agglomerated prior to compaction using organic binders. Cemented carbides are sintered to essentially full density using persistent liquid phases above the eutectic temperature for the composition. Additives are used to inhibit grain growth during sintering (64). After sintering hot isostatic pressing is useful for sealing any residual pores (65). Finally, many cemented carbides are coated with hard surface layers to improve their performance in machining and drilling operations.

The microstructure and composition control the properties of liquid phase sintered cemented carbides. Porosity is a major detriment to the properties and must be maintained below 0.5% (56,66). Also, interface compounds degrade the properties. The amount of matrix, grain separation, and grain size all have recognized effects on properties (67-70). Since the grains are hard and brittle, the cemented carbides are dependent on the matrix for toughness and fracture resistance. The best properties are associated with a strong bond between the matrix and grains. Several relations involving these characteristics are discussed in Chapter 9. A fine grain size with a narrow size distribution is most useful for milling, cutting, and wear applications.

Under proper processing conditions, the sintered materials have properties which exceed that of the component phases. Table 10.4 gives some examples of the properties attained for commercial cemented carbide compositions. The attractive attributes include the high elastic modulus, strength, and hardness. However, because of the low ductility, the cemented carbides have a low fracture toughness and fatigue resistance (7,71).

TABLE 10.3

General Compositions of Cemented Carbides

Carbide	Binder	wt.% Binder
WC	Co	3 to 30
WC+TiC	Co	5 to 20
WC+TaC	Co	5 to 20
WC+VC	Co	5 to 20
TiC	Mo+Ni	10 to 20
WC	Ni	2 to 30
Cr_2C_3	Ni+Co	10 to 20
TiC	Fe	20 to 50

TABLE 10.4

Typical Properties of Industrial Cemented Carbides

Type	wt.% Binder	Density g/cm^3	Elastic Modulus, GPa	Transverse Strength, GPa	Hardness HRA
TiC-Fe	55	6.6	300	1.9	87
WC-Co	14	14.1	520	2.8	87
WC-Co	11	14.4	550	2.7	90
WC-Co	6	15.0	630	1.9	91
WC-Co	3	15.2	680	1.5	93
WC-TaC-TiC-Co	5	12.0	500	1.5	92

Applications for the cemented carbides, and related borides, oxides, and nitrides, are extensive. The attributes of strength, stiffness, hardness, and wear resistance contribute to several uses in machining, cutting, grinding, drilling, and hardfacing (4,9,55,72,73). As a tool material the cemented carbides are used as machining inserts, punches, dies, cutting tools, milling inserts, and saw blades. In wear applications, the cemented carbides are useful because of their resistance to erosion, cavitation, abrasion, and penetration. Thus, they have applications ranging from valve seats to tunneling equipment. Rock drilling equipment relies on cemented carbide drills because of the combined cutting and wear properties. Other uses for cemented carbides include electrical contacts, abrasives for grinding, military projectiles for piercing armor, and substrates for processing semiconductors. Beyond the typical carbide compositions shown in Table 10.4, there are several complex carbide systems, and several new borides or nitrides with similar attractive property combinations.

D. Heavy Alloys

The tungsten heavy alloys are two phase composites formed by persistent liquid phase sintering of mixed powders (74-78). Transition metals such as Ni, Co, Fe, and Cu in various combinations are used to form the liquid phase through eutectic reactions with tungsten. The most popular alloys are based on nickel and iron additions in the ratio of 7:3. Other alloy compositions approximate Ni:Co ratios of 1:1 and Ni:Cu ratios of 2:1. The nickel is important to solubility and wetting during sintering (75,79-82). Typically at least 10% liquid exists during the sintering cycle. This is controlled by the amount of additive, tungsten solubility in the liquid, and sintering temperature. Mixed elemental powders are sintered at temperatures typically 20 to 40 °C over the eutectic temperature for times of 30 to 40 minutes. Cooling from the sintering temperature must be slow to avoid solidification pores and strains in the matrix (83). Additionally, post-sintering heat treatments are used to relieve strains, remove hydrogen, and minimize impurity segregation (78,84-88). The resulting combination of density, strength, ductility, atomic number, melting temperature, and toughness is unique.

The properties of the heavy alloys are sensitive to pores, intermetallic phases, impurities, residual hydrogen, thermal strains, and microstructure. Table 10.5 gives a few examples of the optimal properties possible under closely controlled processing conditions. The fracture path shows dramatic changes due to changes in the processing cycle. The alloys exhibit a classic trade-off between strength and ductility. Fine grain sizes give higher strengths but lower ductilities. Furthermore as the tungsten content increases, the ductility decreases because of an increasing contiguity (89).

The applications for the tungsten heavy alloys typically rely on the combination of high sintered densities and good mechanical properties. (23,90-97). Thus, they are used as radiation shields, counterbalance weights, vibration dampers, gyroscope components, and inertial control devices. Another use is for armor piercing projectiles because of the large kinetic energy associated with the high density. Additionally, tungsten heavy alloys are used to improve the performance of sporting equipment such as golf clubs and darts. High temperature uses include welding rod holders, electrical contacts, die casting tools, extrusion tools, and spark erosion tools for electrode discharge machining. The heavy alloys have high absolute densities, good strength and ductility in a high temperature material. These attributes make them unique alloys.

E. Silicon Nitride Systems

Ceramics based on silicon nitride offer several high temperature options not available with other engineering materials. The silicon nitride based systems provide high temperature oxidation resistance, strength, thermal shock resistance, and creep resistance. Additionally, silicon nitride has a low density, low coefficient of friction, low thermal expansion coefficient, low thermal conductivity, and is wear resistant. However, silicon nitride is a

TABLE 10.5

Properties for Tungsten Heavy Alloys

Composition wt.%	Density g/cm^3	Elastic Modulus, GPa	Strength MPa	Elongation %
90W-7Ni-3Fe	17.1	345	900	29
93W-4.9Ni-2.1Fe	17.7	360	910	30
95W-3.5Ni-1.5Fe	18.2	375	940	32
97W-2.1Ni-0.9Fe	18.6	380	950	19

covalent ceramic and is difficult to consolidate by solid state sintering because of a low diffusivity. As a consequence several liquid phase sintering techniques have evolved for processing to full density. The additives can form either persistent or transient liquid phases (9,98-101). The sintered densities are 98 to 99.9% of theoretical, with actual levels ranging from 3.2 to 3.8 g/cc depending on the composition. The most common additives are Si, MgO, AlN, BeO, Al_2O_3, SiO_2, ZrO_2, Y_2O_3, and CeO_2.

The use of liquid phase sintering balances the easier processing against the potential for degraded high temperature properties. The liquid forms a wetting film along the grain boundaries and appears as a glass phase in some of the compositions (100,102). Although this phase aids densification and helps avoid decomposition of the silicon nitride, it also dictates the high temperature properties (103,98). For this reason transient liquid phase sintering is probably the most useful consolidation technique, especially when combined with an external pressure.

The main applications for liquid phase sintered silicon nitrides are in structural and mechanical components for high temperature use. The low density of silicon nitride coupled with the toughness and resistance to temperature, thermal shock, and oxidation make it an ideal material for high temperature structural applications such as pistons, pumps, turbine components, and bearings. Early uses were in gas turbines for operating temperatures near 1300°C. Subsequent developments include wear components (bushings and bearings) for hostile environments, and as semiconductor packaging materials. Several current development efforts are examining the potential for using silicon nitride in automobile engines. The advantages of high hardness, temperature resistance, light weight, and low friction make these systems ideal candidates for cylinder liners, pistons, valves, nozzles, drawing dies, high speed cutting tools, and furnace components. As cutting tools, the silicon nitrides operate at higher cutting speeds than the cemented carbides.

F. Other Applications

As is evident from Table 10.1, there are several other applications for liquid phase sintering in everyday use. The materials include refractories, alloys, ceramics, cermets, and composites. These materials are used for applications such as grinding wheels, dental amalgams, brakes, clutches, tool steels, nuclear fuels, bearings, cutting wheels, permanent magnets, transformer cores, automotive parts, electrical contacts, capacitors, filters, and welding electrodes. Obviously, this is an extensive list of possible uses and it is beyond the scope of this book to detail all of these applications. However, it is clear that liquid phase sintering is an important industrial process. These several applications make practical use of the basic concepts developed in earlier chapters of this book. It should be recognized that the range of materials results in several different combinations of solubility, diffusivity, wetting, and dihedral angle. These variables combine with the various particle characteristics and processing options to provide the range of attributes needed to satisfy the applications listed in Table 10.1.

G. Summary

The list of applications provided in this chapter illustrates that liquid phase sintering is useful to several technologies. It is an important manufacturing process which allows for the fabrication of unique materials. Sintering in the presence of a liquid phase is rapid and aids the attainment of full density in relatively short sintering cycles. Also, besides faster sintering, a liquid makes improved properties possible.

As is also evident, a liquid allows sintering at lower temperatures than often possible using solid state techniques; thus, it provides processing flexibility. For example, in ferrous alloys furnace design is simplified if the sintering temperature can be kept below 1120°C. However, lower sintering temperatures lead to reduced levels of sintering. With the formation of a liquid phase, a lower sintering temperature is possible, often with improved properties. It is also evident that several materials can only be fabricated by liquid phase sintering; for example such important materials as cemented carbides, tungsten heavy alloys, and dental amalgams.

To understand liquid phase sintering there are several variables that must be considered. From the thermodynamic view, the melting behavior, solubility, and interfacial energies are most important. The interfacial energies dictate the dihedral angle and wetting of the solid by the liquid. The kinetics of liquid phase sintering depend on the spreading, capillarity, and penetration of the liquid. Diffusion of the solid species through the liquid determines the coarsening and intermediate stage densification rates. As was pointed out in Chapter 8, there are several processing factors of great importance to liquid phase sintering. These include the particle size, homogeneity of the additive, sintering time, sintering temperature, green density, and sintering atmosphere.

Together the processing variables and the material characteristics determine the sintered microstructure. The grain size, grain separation, and amount of matrix phase have the greatest effect on properties. Additionally, the contiguity and connectivity of the microstructure affect both the processing and properties. Finally, the properties of materials processed by liquid phase sintering have been indicated in these last two chapters. The range of applications shows that mechanical, physical, nuclear, magnetic, electrical, and thermal properties are of concern for the several industrial products. Furthermore, the behavior of a liquid phase sintered material will depend on the service environment and the material properties. A combination of several of the above factors is important to successful applications. It has been indicated throughout this presentation that actual liquid phase sintered materials have a variety of characteristics. Thus, generalizations are difficult. Likewise the variety available with liquid phase sintering contributes to its widespread use, but also provides a challenge in studying this unique processing technology. This book has provided the diligent reader with a background in the fundamentals of liquid phase sintering in preparation for in-depth study of the various applications.

H. References

1. J. W. Butcher and J. N. Lowe, "Activated Sintering in Beryllium Powders by Selective Addition of Trace Elements," *Beryllium Technology*, vol.1, L. M. Schetky and H. A. Johnson (eds.), Gordon and Breach, New York, NY, 1966, pp.501-522.
2. C. E. Bates and B. R. Patterson, "Transient Liquid Phase Sintering of P/M Titanium Alloys," Report No. 4805-XIV, Southern Research Institute, Birmingham, AL, February 1983.
3. R. Kieffer, G. Jangg, and P. Ettmayer, "Sintered Superalloys," *Powder Met. Inter.*, 1975, vol.7, pp.126-130.
4. F. V. Lenel, *Powder Metallurgy Principles and Applications*, Metal Powder Industries Federation, Princeton, NJ, 1980.
5. G. Dowson, "The Sintering of Bronze," *Metal Powder Rep.*, 1984, vol.39, pp.71-73.
6. E. Peissker, "Pressing and Sintering Characteristics of Powder Mixtures for Sintered Bronze 90/10 Containing Different Amounts of Free Tin," *Modern Developments in Powder Metallurgy*, vol.7, H. H. Hausner and W. E. Smith (eds.), Metal Powder Industries Federation, Princeton, NJ, 1974, pp.597-614.
7. E. A. Almond, "Deformation Characteristics and Mechanical Properties of Hardmetals," *Science of Hard Materials*, R. K. Viswanadham, D. J. Rowcliffe and J. Gurland (eds.), Plenum Press, New York, NY, 1983, pp.517-557.
8. D. Moskowitz and M. Humenik, "Cemented TiC Base Tools with Improved Deformation Resistance," *Modern Developments in Powder Metallurgy*, vol.14, H. H. Hausner, H. W. Antes and G. D. Smith (eds.), Metal Powder Industries Federation, Princeton, NJ, 1981, pp.307-320.
9. M. H. Lewis and R. J. Lumby, "Nitrogen Ceramics: Liquid Phase Sintering," *Powder Met.*, 1983, vol.26, pp.73-81.
10. K. M. Kulkarni, A. Ashurst, and M. Svilar, "Role of Additives in Full Dense Sintering of Tool Steels," *Modern Developments in Powder Metallurgy*, vol.13, H. H. Hausner, H. W. Antes and G. D. Smith (eds.), Metal Powder Industries Federation, Princeton, NJ, 1981, pp.93-120.
11. J. F. Bates and A. G. Knapton, "Metals and Alloys in Dentistry," *Inter. Metals Revs.*, 1977, vol.22, pp.39-60.
12. B. E. Walker, R. W. Rice, R. C. Pohanka, and J. R. Spann, "Densification and Strength of Barium Titanate with LF and MgO Additives," *Bull. Amer. Ceramic Soc.*, 1976, vol.55, pp.274-284.
13. H. U. Anderson and M. C. Proudian, "Sintering of LiF Fluxed Strontium Titanate," *Sintering and Heterogeneous Catalysis*, G. C. Kuczynski (ed.), Plenum Press, New York, NY, 1984, pp.281-292.
14. P. Ayers, "PM Contacts for Contactors and Circuit Breakers," *Metal*

Powder Rep., 1984, vol. ,pp.593-595.
15. G. J. Witter and W. R. Warke, "A Correlation of Material Toughness, Thermal Shock Resistance, and Microstructure of High Tungsten, Silver-Tungsten Composite Materials," *IEEE Trans. Parts Hybrids Packaging*, 1975, vol.11, pp.21-29.
16. A. N. Klein, R. Oberacker, and F. Thummler, "High Strength Si-Mn-Alloyed Sintered Steels," *Powder Met. Inter.*, 1985, vol.17, pp.13-16.
17. S. B. Domsa, D. Topan, and D. Micu, "On the Elaboration and Properties of the High Graphical-Iron Liquid Phase Sintered Materials," *Proceedings Fourth European Symposium for Powder Metallurgy*, vol.3, Societe Francaise de Metallurgie, Grenoble, France, 1975, pp.9-5-1 to 9-5-6.
18. J. N. Brecker, "Analysis of Bond Formation in Vitrified Abrasive Wheels," *J. Amer. Ceramic Soc.*, 1974, vol.57, pp.486-489.
19. F. F. Lange and T. K. Gupta, "Sintering of SiC with Boron Compounds," *J. Amer. Ceramic Soc.*, 1976, vol.59, pp.537-538.
20. N. Reiter and J. Kolaska, "Schneidstoffe, Stand der Technik und Entwicklungstendenzen," *Proceedings Eleventh International Plansee Seminar*, vol.2, H. Bildstein and H. M. Ortner (eds.), Metallwerk Plansee, Reutte, Austria, 1985, pp.335-376.
21. R. Wahling, P. Beiss, and W. J. Huppmann, "Sintering Behaviour and Performance Data of HSS-Components," *Proceedings Sintering Theory and Practice Conference*, The Metals Society, London, UK, 1984, pp.15.1-15.5.
22. P. A. Dearnley and V. Thompson, "An Evaluation of the Failure Mechanisms of Ceramics and Coated Carbides used for Machining Stainless Steels," *Proceedings Eleventh International Plansee Seminar*, vol.2, H. Bildstein and H. M. Ortner (eds.), Metallwerk Plansee, Reutte, Austria, 1985, pp.709-745.
23. D. J. Jones and P. Munnery, "Production of Tungsten Alloy Penetration Radiation Shields," *Powder Met.*, 1967, vol.10, pp.156-173.
24. K. W. Lay, "Grain Growth in Urania-Alumina in the Presence of a Liquid Phase," *J. Amer. Ceramic Soc.*, 1968, vol.51, pp.373-376.
25. P. J. Jorgensen and R. W. Bartlett, "Liquid-Phase Sintering of $SmCo_5$," *J. Appl. Phys.*, 1973, vol.44, pp.2876-2880.
26. Y. G. Bogatin, "Effect of Phase and Structural Transformations Occurring During Liquid-Phase Sintering on the Magnetic Properties of Samarium Cobalt Magnets," *Soviet Powder Met. Metal Ceram.*, 1978, vol.17, pp.393-398.
27. M. L. Green and C. C. Wong, "Liquid Phase Sintering of a Cr-Co-Fe Permanent Magnet Alloy," *Modern Developments in Powder Metallurgy*, vol.12, H. H. Hausner, H. W. Antes and G. D. Smith (eds.), Metal Powder Industries Federation, Princeton, NJ, 1981, pp.453-472.
28. D. L. Johnson and I. B. Cutler, "The Use of Phase Diagrams in the Sintering of Ceramics and Metals," *Phase Diagrams*, vol.2, A. M. Alper (ed.), Academic Press, New York, NY, 1970, pp.265-291.
29. J. White, "Microstructure and Grain Growth in Ceramics in the Presence of a Liquid Phase," *Sintering and Related Phenomena*, G. C. Kuczynski (ed.), Plenum Press, New York, NY, 1973, pp.81-108.
30. B. D. Storozh and P. S. Kislyl, "Sintering of Tungsten-Alumina Cermets in the Presence of a Liquid Phase," *Soviet Powder Met. Metal Ceram.*, 1974, vol.13, pp.712-716.
31. M. Paulus, F. Laher-Lacour, P. Dugleux, and A. Dubon, "Defects and

Transitory Liquid Phase Formation during the Sintering of Mixed Powders," *Trans. J. British Ceram. Soc.*, 1983, vol.82, pp.90-98.
32. W. D. Kingery, H. K. Bowen, and D. R. Uhlmann, *Introduction to Ceramics*, second edition, Wiley-Interscience, New York, NY, 1976.
33. G. Jangg, M. Drozda, H. Danninger, and G. Eder, "Magnetic Properties of Sintered Fe-P Materials," *Powder Met. Inter.*, 1984, vol.16, pp.264-267.
34. G. Jangg, M. Drozda, H. Danninger, H. Wibbeler, and W. Schatt, "Sintering Behavior, Mechanical and Magnetic Properties of Sintered Fe-Si Materials," *Inter. J. Powder Met. Powder Tech.*, 1984, vol.20, pp.287-300.
35. C. Durdaller, "The Effect of Additions of Copper, Nickel and Graphite on the Sintered Properties of Iron-Base Sintered P/M Parts," *Prog. Powder Met.*, 1969, vol.25, pp.73-100.
36. R. M. German and K. A. D'Angelo, "Enhanced Sintering Treatments for Ferrous Powders," *Inter. Metals Rev.*, 1984, vol.29, pp.249-272.
37. M. R. Pickus, "Improving the Properties of P/M Steels Through Liquid Phase Sintering," *Inter. J. Powder Met. Powder Tech.*, 1984, vol.20, pp.311-323.
38. P. J. McGinn, A. E. Miller, P. Kumar, and A. J. Hickl, "Mechanisms of Liquid Phase Sintering in Stellite Alloy No. 6 PM," *Prog. Powder Met.*, 1982, vol.38, pp.449-462.
39. J. L. Ellis, "Wear Resistant Alloy Bonded Carbides Produced by PM," *Powder Met. Inter.*, 1984, vol.16, pp.53-55.
40. A. Ball and A. W. Paterson, "Microstructural Design of Erosion Resistant Hard Materials," *Proceedings Eleventh International Plansee Seminar*, vol.2, H. Bildstein and H. M. Ortner (eds.), Metallwerk Plansee, Reutte, Austria, 1985, pp.377-391.
41. B. A. James, "Liquid Phase Sintering in Ferrous Powder Metallurgy," *Proceedings Sintering Theory and Practice Conference*, The Metals Society, London, UK, 1984, pp.12.1-12.14.
42. R. M. German, "An Overview of Enhanced Sintering Treatments for Iron," *Sintering and Heterogeneous Catalysis*, G. C. Kuczynski, A. E. Miller, and G. A. Sargent (eds.), Plenum Press, New York, NY, 1984, pp.103-114.
43. B. Loy and R. J. Dower, "The Effect of Boron on some Properties of Sintered Iron-Carbon Alloys," *Proceedings P/M-82*, Associazione Italiana di Metallurgia, Milan, Italy, 1982, pp.307-314.
44. J. B. Long and D. A. Robins, "Improving the Sintering Performance of Iron Powder by the Addition of Tin," *Modern Developments in Powder Metallurgy*, vol.4, H. H. Hausner (ed.), Plenum Press, New York, NY, 1971, pp.303-313.
45. D. J. Lee and R. M. German, "Sintering Behavior of Iron-Aluminum Powder Mixes," *Inter. J. Powder Met. Powder Tech.*, 1985, vol.21, pp.9-21.
46. B. F. Kieback, W. Schatt, and G. Jangg, "Titanium-Alloyed Sintered Steels," *Powder Met. Inter.*, 1984, vol.16, pp.207-212.
47. W. J. Huppmann, W. Kaysser, and F. J. Esper, "Sintering of Iron-Manganese with Tin Additions," *Proceedings Fourth European Symposium for Powder Metallurgy*, vol.1, Societe Francaise de Metallurgie, Grenoble, France, 1975, pp.3-7-1 to 3-7-6.
48. M. D. Hamiuddin and G. S. Upadhyaya, "Effect of Copper on Sintering of Phosphorus Containing Iron," *Trans. Powder Met. Assoc. India*, 1979, vol.6, pp.57-63.

49. R. Duckett and D. A. Robins, "Tin Additions to Aid the Sintering of Iron Powder," *Metallurgia*, 1966, (Oct.), pp.163-167.
50. G. Matsumura, "Sintering of Iron Wires with a Liquid Phase," *Inter. J. Powder Met.*, 1969, vol.5, no.2, pp.55-61.
51. T. Krantz, "Effect of Density and Composition on the Dimensional Stability and Strength of Iron-Copper Alloys," *Inter. J. Powder Met.*, 1969, vol.5, no.3, pp.35-43.
52. J. Gurland, "Observations on the Structure and Sintering Mechanism of Cemented Carbides," *Trans. TMS-AIME*, 1959, vol.215, pp.601-608.
53. J. Gurland and J. T. Norton, "Role of the Binder Phase in Cemented Tungsten Carbide-Cobalt Alloys," *Trans. AIME*, 1952, vol.194, pp.1051-1056.
54. R. F. Snowball and D. R. Milner, "Densification Processes in the Tungsten Carbide-Cobalt System," *Powder Met.*, 1968, vol.11, pp.23-40.
55. P. Schwarzkopf and R. Kieffer, *Cemented Carbides*, Macmillan, New York, NY, 1960.
56. H. E. Exner and J. Gurland, "A Review of Parameters Influencing some Mechanical Properties of Tungsten Carbide-Cobalt Alloys," *Powder Met.*, 1970, vol.13, pp.13-31.
57. R. Warren and M. B. Waldron, "Microstructural Development During the Liquid-Phase Sintering of Cemented Carbides II. Carbide Grain Growth," *Powder Met.*, 1972, vol.15, pp.180-201.
58. R. Kieffer, "Theoretical Aspects of Sintering of Carbides," *The Physics of Powder Metallurgy*, W. E. Kingston (ed.), McGraw-Hill, New York, NY, 1951, pp.278-291.
59. R. J. Nelson and D. R. Milner, "Liquid-Flow Densification in the Tungsten Carbide-Copper System," *Powder Met.*, 1971, vol.14, pp.39-63.
60. R. J. Nelson and D. R. Milner, "Densification Processes in the Tungsten Carbide-Cobalt System," *Powder Met.*, 1972, vol.15, pp.346-363.
61. R. Warren and M. B. Waldron, "Microstructural Development During the Liquid-Phase Sintering of Cemented Carbides I. Wettability and Grain Contact," *Powder Met.*, 1972, vol.15, pp.166-180.
62. W. May, "Phase Decomposition and Grain Growth in (W,Ti)C-Co-Alloys," *J. Mater. Sci.*, 1971, vol.6, pp.1209-1213.
63. M. Fukuhara and H. Mitani, "Mechanisms of Grain Growth in Ti(C,N)-Ni Sintered Alloys," *Powder Met.*, 1982, vol.25, pp.62-68.
64. D. Y. Kim and A. Accary, "Mechanisms of Grain Growth Inhibition During Sintering of WC-Co Based Hard Metals," *Sintering Processes*, G. C. Kuczynski (ed.), Plenum Press, New York, NY, 1980, pp.235-244.
65. S. Amberg, E. A. Nylander, and B. Uhrenius, "The Influence of Hot Isostatic Pressing on the Porosity of Cemented Carbide," *Powder Met. Inter.*, 1974, vol.6, pp.178-180.
66. L. LeRoux, "Microstructure and Transverse Rupture Strength of Cemented Carbides," *Inter. J. Refract. Hard Met.*, 1984, vol.3, pp.99-100.
67. J. Gurland and P. Bardzil, "Relation of Strength, Composition, and Grain Size of Sintered WC-Co Alloys," *Trans. TMS-AIME*, 1955, vol.203, pp.311-315.
68. J. L. Chermant, A. Deschanvres, and F. Osterstock, "Factors Influencing the Rupture Stress of Hardmetals," *Powder Met.*, 1977, vol.20, pp.63-69.
69. J. L. Chermant and F. Osterstock, "Elastic and Plastic Characteristics of WC-Co Composite Materials," *Powder Met. Inter.*, 1979, vol.11, pp.106-109.

70. J. L. Chermant, M. Coster, G. Hautier, and P. Schaufelberger, "Statistical Analysis of the Behaviour of Cemented Carbides under High Pressure," *Powder Met.*, 1974, vol.17, pp.85-102.
71. M. J. Murray, "Fracture of WC-Co Alloys: An Example of Spatially Constrained Crack Tip Opening Displacement," *Proc. Royal Soc. Lond. A*, 1977, vol.A356, pp.483-508.
72. R. Kieffer and F. Benesovsky, *Hartmetalle*, Springer-Verlag, Vienna, Austria, 1965.
73. K. J. A. Brookes, *World Directory and Handbook of Hardmetals*, third edition, Engineers' Digest and International Carbide Data, London, UK, 1982.
74. E. C. Green, D. J. Jones, and W. R. Pitkin, "Developments in High-Density Alloys," *Symposium on Powder Metallurgy*, Special Report 58, Iron and Steel Institute, London, UK, 1956, pp.253-256.
75. G. H. S. Price, C. J. Smithells, and S. V. Williams, "Sintered Alloys. Part I - Copper-Nickel-Tungsten Alloys Sintered with a Liquid Phase Present," *J. Inst. Metals*, 1938, vol.62, pp.239-264.
76. E. G. Zukas and H. Sheinberg, "Sintering Mechanisms in the 95% W - 3.5% Ni - 1.5% Fe Composite," *Powder Tech.*, 1976, vol.13, pp.85-96.
77. E. Ariel, J. Barta, and D. Brandon, "Preparation and Properties of Heavy Alloys," *Powder Met. Inter.*, 1973, vol.5, pp.126-129.
78. R. M. German and J. E. Hanafee, "Processing Effects on Toughness for Liquid Phase Sintered W-Ni-Fe," *Processing of Metal and Ceramic Powders*, R. M. German and K. W. Lay (eds.), The Metallurgical Society, Warrendale, PA, 1982, pp.267-282.
79. H. H. Hausner, "Some Modified Heavy Metal Alloys," *Metals and Alloys*, 1943, vol.18, December, pp.437-440.
80. C. J. Li and R. M. German, "Enhanced Sintering of Tungsten - Phase Equilibria Effects on Properties," *Inter. J. Powder Met. Powder Tech.*, 1984, vol.20, pp.149-162.
81. R. H. Krock and L. A. Shepard, "Mechanical Behavior of the Two-Phase Composite, Tungsten-Nickel-Iron," *Trans. TMS-AIME*, 1963, vol.227, pp.1127-1134.
82. H. Takeuchi, "Ductility and Matrix Constitution of Sintered W-Ni-Fe Alloys," *Nippon Kinzoku Gakkaishi*, 1967, vol.31, pp.1064-1069.
83. B. Lux, W. J. Huppmann, G. Jangg, H. Danninger, and W. Pisan, "The Influence of Impurities in Tungsten and Matrix Composition on the Tungsten-Matrix Interfacial Properties of Heavy Metals," Final Report DAJA-80-C-0008, Institut fur Chemische Technologie Anorganischer Stoffe, Technical Universitat Wein, Vienna, Austria, October 1982.
84. M. R. Eisenmann and R. M. German, "Factors Influencing Ductility and Fracture Strength in Tungsten Heavy Alloys," *Inter. J. Refract. Hard Met.*, 1984, vol.3, pp.86-91.
85. T. K. Kang, E. T. Henig, W. A. Kaysser, and G. Petzow, "Effect of Cooling Rate on the Microstructure of a 90W-7Ni-3Fe Heavy Alloy," *Modern Developments in Powder Metallurgy*, vol.14, H. H. Hausner, H. W. Antes and G. D. Smith (eds.), Metal Powder Industries Federation, Princeton, NJ, 1981, pp.189-203.
86. B. C. Muddle and D. V. Edmonds, "Interfacial Segregation and Embrittlement in Liquid Phase Sintered Tungsten Alloys," *Metal Sci.*, 1983, vol.17, pp.209-218.
87. L. Ekbom, "The Influence of Microstructure of Liquid-Sintered Tungsten-Base Composites on the Mechanical Properties," *Scand. J. Metall.*, 1976, vol.5, pp.179-184.

88. C. C. Ge, X. I. Xia, and E. T. Henig, "Effects of some Sintering Parameters and Vacuum Heat Treatment on the Mechanical Properties and Fracture Mode of Heavy Alloys," *Proceedings P/M-82*, Associazione Italiana di Metallurgia, Milan, Italy, 1982, pp.709-714.
89. R. M. German, "Microstructure Limitations of High Tungsten Content Heavy Alloys," *Proceedings Eleventh International Plansee Seminar*, vol.1, H. Bildstein and H. M. Ortner (eds.), Metallwerk Plansee, Reutte, Austria, 1985, pp.143-161.
90. B. Nathan, "Many Options for Sintered Tungsten Alloys," *Metal Powder Rep.*, 1985, vol.40, pp.271-276.
91. E. I. Larson and P. C. Murphy, "Characteristics and Applications of High Density Tungsten-Based Composites," *Can. Min. Met. Bull.*, 1965, April, pp.413-420.
92. N. C. Kothari, "Factors Affecting Tungsten-Copper and Tungsten-Silver Electrical Contact Materials," *Powder Met. Inter.*, 1982, vol.14, pp.139-159.
93. I. H. Moon and J. S. Lee, "Sintering of W-Cu Contact Materials with Ni and Co Dopants," *Powder Met. Inter.*, 1977, vol.9, pp.23-24.
94. I. H. Moon and J. S. Lee, "Activated Sintering of Tungsten-Copper Contact Materials," *Powder Met.*, 1979, vol.22, pp.5-7.
95. P. Walkden, J. N. Albiston, and F. R. Sale, "Reduction of Tungstates for Production of Silver-Tungsten and Silver-Tungsten-Nickel Electrical Contacts," *Powder Met.*, 1985, vol.28, pp.36-42.
96. G. H. Gessinger and K. N. Melton, "Burn-off Behaviour of W-Cu Contact Materials in an Electric Arc," *Powder Met. Inter.*, 1977, vol.9, pp.67-72.
97. I. H. Moon and W. J. Huppmann, "Sintering Behavior of Tungsten-Silver Contact Materials with Cobalt Additions," *Powder Met. Inter.*, 1974, vol.6, pp.190-194.
98. J. Lorenz, J. Weiss, and G. Petzow, "Dense Silicon Nitride Alloys: Phase Relations and Consolidation, Microstructure and Properties," *Advances in Powder Technology*, G. Y. Chin (ed.), American Society for Metals, Metals Park, OH, 1982, pp.289-308.
99. G. R. Terwilliger and F. F. Lange, "Pressureless Sintering of Si_3N_4," *J. Mater. Sci.*, 1975, vol.10, pp.1169-1174.
100. J. Mukerji, P. Greil, and G. Petzow, "Sintering of Silicon Nitride with a Nitrogen Rich Liquid Phase," *Sci. Sintering*, 1983, vol.15, pp.43-53.
101. S. Baik and R. Raj, "Effect of Silicon Activity on Liquid-Phase Sintering of Nitrogen Ceramics," *J. Amer. Ceramic Soc.*, 1985, vol.68, pp.C124-C126.
102. L. K. V. Lou, T. E. Mitchell, and A. H. Heuer, "Impurity Phases in Hot Pressed Silicon Nitride," *J. Amer. Ceramic Soc.*, 1984, vol.67, pp.392-396.
103. J. E. Marion, A. G. Evans, M. D. Drory, and D. R. Clarke, "High Temperature Failure Initiation in Liquid Phase Sintered Materials," *Acta Met.*, 1983, vol.31, pp.1445-1457.

INDEX

GPSR Compliance

The European Union's (EU) General Product Safety Regulation (GPSR) is a set of rules that requires consumer products to be safe and our obligations to ensure this.

If you have any concerns about our products, you can contact us on ProductSafety@springernature.com

In case Publisher is established outside the EU, the EU authorized representative is:

Springer Nature Customer Service Center GmbH
Europaplatz 3
69115 Heidelberg, Germany

Batch number: 10761599

Printed by Printforce, the Netherlands